Willem-Jan Zwanenburg

Degradation Processes of Railway Switches & Crossings

Willem-Jan Zwanenburg

Degradation Processes of Railway Switches & Crossings

To Improve Maintenance & Renewal Planning on the Swiss Railway Network

Südwestdeutscher Verlag für Hochschulschriften

Impressum/Imprint (nur für Deutschland/ only for Germany)
Bibliografische Information der Deutschen Nationalbibliothek: Die Deutsche Nationalbibliothek verzeichnet diese Publikation in der Deutschen Nationalbibliografie; detaillierte bibliografische Daten sind im Internet über http://dnb.d-nb.de abrufbar.
Alle in diesem Buch genannten Marken und Produktnamen unterliegen warenzeichen-, marken- oder patentrechtlichem Schutz bzw. sind Warenzeichen oder eingetragene Warenzeichen der jeweiligen Inhaber. Die Wiedergabe von Marken, Produktnamen, Gebrauchsnamen, Handelsnamen, Warenbezeichnungen u.s.w. in diesem Werk berechtigt auch ohne besondere Kennzeichnung nicht zu der Annahme, dass solche Namen im Sinne der Warenzeichen- und Markenschutzgesetzgebung als frei zu betrachten wären und daher von jedermann benutzt werden dürften.

Verlag: Südwestdeutscher Verlag für Hochschulschriften Aktiengesellschaft & Co. KG
Dudweiler Landstr. 99, 66123 Saarbrücken, Deutschland
Telefon +49 681 37 20 271-1, Telefax +49 681 37 20 271-0, Email: info@svh-verlag.de
Zugl.: Lausanne, École Polytechnique Fédérale de Lausanne, Diss., 2009

Herstellung in Deutschland:
Schaltungsdienst Lange o.H.G., Berlin
Books on Demand GmbH, Norderstedt
Reha GmbH, Saarbrücken
Amazon Distribution GmbH, Leipzig
ISBN: 978-3-8381-0516-1

Imprint (only for USA, GB)
Bibliographic information published by the Deutsche Nationalbibliothek: The Deutsche Nationalbibliothek lists this publication in the Deutsche Nationalbibliografie; detailed bibliographic data are available in the Internet at http://dnb.d-nb.de.
Any brand names and product names mentioned in this book are subject to trademark, brand or patent protection and are trademarks or registered trademarks of their respective holders. The use of brand names, product names, common names, trade names, product descriptions etc. even without a particular marking in this works is in no way to be construed to mean that such names may be regarded as unrestricted in respect of trademark and brand protection legislation and could thus be used by anyone.

Publisher:
Südwestdeutscher Verlag für Hochschulschriften Aktiengesellschaft & Co. KG
Dudweiler Landstr. 99, 66123 Saarbrücken, Germany
Phone +49 681 37 20 271-1, Fax +49 681 37 20 271-0, Email: info@svh-verlag.de

Copyright © 2009 by the author and Südwestdeutscher Verlag für Hochschulschriften Aktiengesellschaft & Co. KG and licensors
All rights reserved. Saarbrücken 2009

Printed in the U.S.A.
Printed in the U.K. by (see last page)
ISBN: 978-3-8381-0516-1

*In remembrance of Professor Robert Rivier
for creating the opportunity*

Voor Bas en Anouk

Acknowledgement

There is a large amount of people who contributed to this thesis – sometimes even without knowing it themselves that they were. The first and perhaps most important is unfortunately not among us anymore. Professor Robert Rivier, who passed away the 10th of May 2007 at the too young age of 64, gave me the chance and created the possibility to fulfil a dream, gain unique experiences and see something of the world. From the first time we met, on the station of Lausanne, we were talking about the same thing in full understanding: railways. These conversations luckily have not stopped and I am sure it would not have stopped if it was not for that sad day last year.

It was Professor André-Gilles Dumont who was willing to take over the task of guiding me through the last bits of this thesis, for which I am very thankful. Beside this there are the members of the jury to which I want to express my gratitude: Professor Eugen Brühwiler and Professor Anton Schleiss from the EPFL, Professor Ulrich Weidmann from ETH Zürich and Dr. Clive Roberts from the University of Birmingham.

As major contributors from outside the EPFL to fulfilling this thesis can be listed: Markus Bertschy, Claude Chevalley, Fritz Held, Christoph Hofmann, Arnold Kellenberger, Daniel Kuster, Heinz Rickli and Michel Thomet from SBB for supplying expert input and the databases needed for this study; Bruno Magoni for helping with the analyses of the databases; Carolina Meier-Hirmer from SNCF for providing a welcome insight in statistics.

I could not have finished this study, without the lunches, coffee breaks, *apéro's*, Christmas dinners and much more in- and outside our nice work environment with current and former colleagues from our great lab: Anne, Chiara, Christine, Isabelle, Daniel, Jean-Daniel, Jean-Pierre, Panos, Raphaëlle, Regi, Susana, Yves, Zachary and everyone I might have forgotten including Master and Bachelor students from the EPFL and visiting students, during the last 5 years.

What rests are all the friends in the Netherlands, in Switzerland or anywhere else who supported me during the last 5 years and with whom I had loads of pleasure. I hope you don't mind that I cannot name all of you here, but that you feel how much you all meant and mean to me.

The last part of this acknowledgement is reserved for my family: my parents, who supported all the crazy plans I have carried out until now, Jan-Bart, Daniëlle and Bas & Anouk – without whom I could not play the role of nice-uncle-in-a-foreign-country; the rest of my family: uncles, aunts, cousins. I thank you all.

Lausanne, 14th of July 2008

"Wear is the result of friction, in all relationships"

Abstract

Approximately 25 percent of the budget for the maintenance and renewal of railway tracks –in Switzerland more than a billion Swiss Franks– is used for the switches (points) and crossings (S&C). While in the mean time the budget expenditure for the maintenance and renewal of plain track is optimized with e.g. the help of decision support systems, for S&C this optimisation stays rather rudimentary. One of the identified causes thereof is the lack of insight in the degradation and deterioration process of S&C.

This study therefore combined several databases of the Swiss Federal Railways (SBB CFF FFS). Statistical analyses are carried out on them to retrieve the lifetime expectancy of complete railway switches (points) & crossings and their respective components, e.g. point rails, stock rails, frog. The expected lifetimes are attributed to different parameters which influence the speed of geometrical degradation or wear of the material, e.g. total train loads (expressed in cumulative tonnages), axle loads, the main direction of the trains, the speed and the quality of the foundation.

Key words: railways, switches and crossings, maintenance and renewal, maintenance optimisation, maintenance planning, maintenance policy, infrastructure management, asset management, degradation, wear

Résumé

Environ 25 pourcents du budget pour l'entretien et le renouvellement des voies ferrées – en Suisse plus de 1 milliard de francs suisses – est utilisé pour l'entretien des aiguillages et des appareils de voie. Pendant que les dépenses consacrées à l'entretien et au renouvellement de la pleine voie sont optimisées, p. ex. à l'aide de systèmes décisionnels, l'optimisation des dépenses pour les appareils de voie reste plutôt rudimentaire. Une des raisons expliquant cela est le manque de connaissance au sujet du processus de dégradation et de l'usure des appareils de voie.

Cette étude combine pour cette raison plusieurs bases de données des Chemins de Fer Fédéraux Suisses (SBB CFF FFS). Des analyses statistiques sont réalisées afin d'obtenir l'espérance de durée de vie des appareils de voie et de ses composants, p.ex. les lames, les sommiers, les contre-rails et les cœurs. Les durées de vie sont liées aux différents paramètres influençant la vitesse de la dégradation géométrique de la voie ou l'usure du matériel, p. ex. la charge cumulée des trains, les charges par essieux, la direction principale des trains sur l'appareil de voie, la vitesse et la qualité de la fondation.

Mots clés: chemin de fer, appareils de voie, aiguillages, entretien et renouvellement, gestion de la maintenance, politique de maintenance, optimisation de maintenance, dégradation, usure

Zusammenfassung

Ungefähr 25 Prozent des Etats für Instandhaltung und Erneuerung von Bahngleisen – und somit in der Schweiz über eine Milliarde Schweizer Franken – werden für Weichen und Kreuzungen aufgewendet. Während mittlerweile der Kostenaufwand für die Instandhaltung und Erneuerung auf der freien Strecke unter anderem mit Hilfe von Entscheidungsunterstützenden Systemen optimiert wurde, sind diese Verbesserungsmöglichkeiten für Weichen und Kreuzungen nach wie vor rudimentär. Einer der vorsächlichen Gründe besteht in diesem Zusammenhang in der fehlenden Einschätzbarkeit des Abnutzungs- und Verschleißgrades von Weichen und Kreuzungen.

Diese Studie verbindet somit verschiedene Datenbanken der Schweizerischen Bundesbahn (SBB CFF FFS). Anhand statistischer Analysen wird die Lebenserwartung kompletter Weichen- und Kreuzungsanlagen und ihrer zugehörigen Komponenten wie Weichenzungen, Backenschienen und einfachen Herzstücken untersucht. Die erwarteten Lebenszeiten werden an verschiedenen Parametern beigemessen welche die geometrische Abschreibung des Materials beeinflussen, z.B. das gesamte Wagenzuggewicht (in kumulativen Tonnen), die Achsladungen, die Hauptzugrichtung, die Geschwindigkeit und die Qualität des Untergrundes.

Schlagworte: Eisenbahnen, Weichen und Kreuzungen, Instandhaltung und Erneuerung, Instandhaltungsoptimierung, Instandhaltungsplanung, Instandhaltungspolitik, Instandhaltungsmanagement, Bestandsmanagement, Abschreibung, Abnutzung

Samenvatting

Ongeveer 25 procent van het budget van meer dan een miljard Zwitserse Frank, voor onderhoud en vernieuwing van spoorwegen wordt besteed aan het onderhouden van wissels en kruisingen. Terwijl al geruime tijd de bestedingen voor het onderhoud van normaal spoor (bogen en rechtstanden) zijn geoptimaliseerd bijvoorbeeld door het gebruik van beslissingsondersteunende computersystemen, is de optimalisatie voor wissels en kruisingen achtergebleven. Een van de geïdentificeerde oorzaken is het gebrek aan inzicht betreffende de processen die ten grondslag liggen aan de geometrische verslechtering of slijtageverschijnselen.

Deze studie combineert daarom verscheidene databanken van de Zwitserse spoorwegen (SBB CFF FFS) waarop analyses werden uitgevoerd om de verwachte levensduur van complete wissels en kruisingen en hun respectievelijke onderdelen, bijvoorbeeld tong en aanslagspoorstaaf, strijkregel of puntstuk, te bepalen. Voor de verwachte levensduur wordt een relatie gelegd met factoren die daar invloed op kunnen uitoefenen, zoals de belasting (uitgedrukt in cumulatieve tonnages), aslasten, de richting van de meeste treinen, de treinsnelheid, de bodemkwaliteit en het type wissel.

Trefwoorden: spoorwegen, wissels en kruisingen, onderhoud en vernieuwing, onderhoudsoptimalisatie, onderhoudsplanning, beheer en onderhoud, onderhoudspolitiek, slijtage

Table of contents

Acknowledgement i
Abstract iii
Résumé iii
Zusammenfassung iv
Samenvatting iv
List of figures vii
List of tables ix
Abbreviations xi
1. Introduction, objectives and limitations 1
   1.1 The necessity of switches & crossings 1
   1.2 The reasons for this study 3
   1.3 Derived objectives of this study 7
   1.4 Terminology 7
2. Description of switches & crossings 9
   2.1 Introduction 9
   2.2 Categorisation by means of curve radius or switch angle 10
   2.3 Categorisation by means of type 11
      2.3.1 Standard turnout 12
      2.3.2 Symmetrical turnout 13
      2.3.3 Combined turnout 14
      2.3.4 Diamond crossing 15
      2.3.5 Diamond crossing with single slip 15
      2.3.6 Diamond crossing with double slips 16
      2.3.7 Special combinations 16
      2.3.8 Special geometrical property: curved switches 17
   2.4 Components of switches & crossings 18
      2.4.1 Switch blades and accompanying stock rail 19
      2.4.2 Frog 21
      2.4.3 Check rail 24
      2.4.4 Common crossing 25
      2.4.5 Intermediate rails 26
      2.4.6 Sleepers 26
      2.4.7 Ballast 27
   2.5 Switches & crossings on non-ballasted tracks 27
   2.6 Synthesis 27
3. Degradation and wear of railway tracks 29
   3.1 Introduction 29
   3.2 Theoretical basis: Forces and loads on railway tracks and S&C in particular 30
      3.2.1 Static loads 30
      3.2.2 Vertical dynamic loads 31
      3.2.3 Horizontal dynamic load 34
      3.2.4 Static and dynamic loads on switches and crossings 34
      3.2.5 Loads vs. degradation of geometry & track component wear 35
   3.3 Modelling the degradation and wear: the approach 37
      3.3.1 Engineering approach 37
      3.3.2 Statistical approach 39
   3.4 Geometrical degradation 39
      3.4.1 Inspection of geometrical quality 40
      3.4.2 State-of-the-art on geometrical degradation modelling 41
   3.5 Tear, wear and plastic deformation of components 44
      3.5.1 Inspection of tear, wear and plastic deformation of components 45
      3.5.2 State-of-the-art on modelling of the tear, wear and plastic deformation 46
   3.6 Reflection on degradation and tear, wear, plastic deformation and inspection of S&C 47
   3.7 Synthesis 48
4. Maintenance and renewal of switches and crossings 51
   4.1 Introduction 51
   4.2 Maintenance actions on S&C 52
   4.3 Renewal actions on S&C 53

4.4		The optimal moment to carry out maintenance or renewal	54
4.5		Maintenance and renewal policy	55
	4.5.1	Description	55
	4.5.2	Life expectancy and substance	55
4.6		Maintenance and renewal (policy) optimisation	56
	4.6.1	Introduction	56
	4.6.2	State of the art on maintenance optimisation	58
4.7		Synthesis	61
5.		Methodology & theoretical basis	63
5.1		Available data & database description	63
	5.1.1	Switches and crossings database, general description	63
	5.1.2	Maintenance works on S&C database	64
	5.1.3	Renewals on S&C database	65
	5.1.4	Loads on the SBB network database	65
	5.1.5	Other databases	65
5.2		The search fro a model: generalities	65
5.3		The Markov-chain approach to the degradation and wear description	66
5.4		The reliability model for the degradation and wear description	66
	5.4.1	Single parameter analysis	68
	5.4.2	Multi-parameter analyses	68
	5.4.3	Parametrical analysis	69
5.5		Testing of the degradation model and introduction of a maintenance policy parameter	70
5.6		Integration in a Life Cycle Cost model	70
5.7		Final remarks on the methodology	70
6.		Modelling of degradation & restoration process	73
6.1		Data handling and database combination	73
	6.1.1	Attributing the load	73
	6.1.2	Attributing the soil quality	74
	6.1.3	The position of the S&C in Switzerland	74
6.2		Applied filters	74
6.3		Presumptions	75
6.4		Complete switch or crossing renewal	76
	6.4.1	Replacement ages of complete switches and crossings during the years	76
6.5		Complete switch or crossing replacement modelling	78
	6.5.1	Data analysis	78
	6.5.2	Results for a standard turnout (EW)	79
	6.5.2.1	Single parameter analysis for EW	79
	6.5.2.2	Multi-parameter analysis for EW	88
	6.5.3	Results for a diamond crossing with double slips (DKW)	89
	6.5.3.1	Single parameter analysis for DKW	90
	6.5.3.2	Multi-parameter analysis for DKW	92
6.6		Synthesis	93
7.		Model testing	95
7.1		Determination of renewal needs for standard turnouts	95
7.2		Reflections on the obtained model	96
7.3		Synthesis	97
8.		Conclusion	99
8.1		General conclusion	99
8.2		The parameters that cause wear and degradation	99
8.3		The results of the multi-parameter model	99
8.4		The added value of this thesis	100
9.		Recommendations	101
9.1		Inspection of switches & crossings	101
9.2		Design of switches & crossings	102
9.3		Maintenance of switches & crossings	102
9.4		Renewal of switches & crossings	102
9.5		Research of switches & crossings	103
10.		References and literature	105
10.1		General	105
10.2		SBB regulations	109
11.		Annexes	111

List of figures

Figure 1 Generalized situation of the use of switches and crossings...1
Figure 2 A train wheelset and the different track gauges and clearances..2
Figure 3 Example of a discontinuous way of changing tracks: a turntable in Muttenz (CH)....................2
Figure 4 Example of a discontinuous way of changing tracks: a travelling platform in Rotterdam (NL)..........3
Figure 5 An expansion switch, Lyon (F) 3
Figure 6 Organisational model, Putallaz (2007)..4
Figure 7 Categorisation of S&C types by their function..11
Figure 8 Directions at a standard right-hand turnout..12
Figure 9 Switch actuator for moving the switch blades, Rotterdam (NL)..13
Figure 10 Symmetrical turnouts, marshalling yard of Villeneuve (F)...14
Figure 11 Combined turnout, marshalling yard of Sibelin (F)...15
Figure 12 Diamond crossing in the junction of St Benoit (F)..15
Figure 13 Diamond crossing with single slip, Zwolle (NL)...16
Figure 14 Diamond crossing with double slips, Rotterdam IJsselmonde (NL)..16
Figure 15 Schematic cross-over with two right-hand turnouts..16
Figure 16 Schematic double intersecting cross-over with two right-hand, two left-hand turnouts and a crossing 17
Figure 17 Double intersecting cross-over at Rotterdam (NL)...17
Figure 18 Almost right-angle crossings at Utrecht Blauwkapel (NL)...17
Figure 19 Left: Special curved diamond crossing with single slip in Bern (CH). Right: Curved standard left hand turnout in Palézieux (CH)..18
Figure 20 Standard right-hand turnout with indication of the important components 19
Figure 21 Schematic view of the support of the wheel by the stock rail and in R6 the switch blade....20
Figure 22 Change from high-rail profile to low rail profile..20
Figure 23 Different cross sections of a high-profile switch blade...21
Figure 24 Left: constructed frog type 1 in Rotterdam (NL); right: cast manganese frog in France......22
Figure 25   Schematic view of the passage of a wheel through the frog including clearances and gauges 23
Figure 26 Left: Swing nose crossing, Amsterdam (NL) and Wanzwil (CH) 24
Figure 27 Check rail on the left, opposite of the frog on the right, Rotterdam (NL)............................24
Figure 28 Cast manganese common crossing, Junction of St. Benoit (F) 25
Figure 29 Schematic representation of a wheel passing a common crossing..25
Figure 30 Moveable switch blades in the common crossing part of a diamond crossing, Kijfhoek (NL)...........26
Figure 31 Possible ways of quality decrease...30
Figure 32 Evolution of the dynamic impact factors for vertical wheel loads acting on the rail [Zicha (1989)] ...32
Figure 33 Relationship between train loads, track geometry and the track materials 36
Figure 34 Train-track longitudinal model..37
Figure 35 Train-track transversal model 38
Figure 36 The way in which plain track geometry is expressed..40
Figure 37 Example of a degradation curve..42
Figure 38 Settlement behaviour of 2 left-hand switches as shown above from Lichtberger (2005).....43
Figure 39 Examples of rail wear. Left: outer rail in a curve. Right: plastic deformation due to overload............44
Figure 40 Example of rail wear: short-pitch corrugation...44
Figure 41 Examples of rail wear. From left to right: Gauge corner defect, running surface defect and head-checks (last one: junction of St. Benoit (F))..45
Figure 42 Relationship between track geometry, track materials and the maintenance and renewal works.........51
Figure 43 Quality during the life of a S&C or its component...54
Figure 44 Quality during the life of a S&C or its component...59
Figure 45 Renewal or maintenance points, from Meier-Hirmer (2007)..67
Figure 46 Map of Switzerland indicating the place of the switches and crossings subject of this study...............74
Figure 47 Average replacement age for simple turnouts in HG1 tracks (1988-2005)............................76
Figure 48 Average replacement age for simple turnouts in HG2 tracks (1988-2005).............................76
Figure 49 Average replacement age for simple turnouts in HG3 tracks (1988-2005).............................77
Figure 50 Average replacement age for double diamond crossing with slips in HG1 tracks (1988-2005).........77
Figure 51 Cumulative lifetime distribution functions of standard turnout replacements in years (left) and tonnage (right) [DfA, SBB]..80

Figure 52 Cumulative lifetime distribution functions of standard turnout replacements related to the turnout angle [DfA, SBB]..82
Figure 53 Cumulative lifetime distribution functions of standard turnout replacements related to the turnout radius [DfA, SBB] 83
Figure 54 On the left a curved turnout (both directions are in a left turn), on the right a top view on a normal turnout with one straight direction and one curved direction..83
Figure 55 Cumulative lifetime distribution functions of curved and straight standard turnouts [DfA, SBB]......84
Figure 56 Cumulative lifetime distribution functions related to the percentage of freight trains [DfA, SBB].....85
Figure 57 Cumulative lifetime distribution functions related to the soil quality [DfA, SBB]...............................86
Figure 58 Cumulative lifetime distribution functions related to the main direction a standard turnout is used [DfA, SBB] 87
Figure 59 Cumulative lifetime distribution functions related to the percentage of freight trains [DfA, SBB].....87
Figure 60 Cumulative lifetime distribution functions of diamond crossings with double slips replacements in years and tonnage [DfA, SBB] 90
Figure 61 Cumulative lifetime distribution functions related to the switch radius/frog angle [DfA, SBB]..........90
Figure 62 Cumulative lifetime distribution functions related to the percentage of freight trains [DfA, SBB].....91
Figure 63 Cumulative lifetime distribution functions related to the percentage of freight trains [DfA, SBB].....92
Figure 64 Renewals per year as registered in the database WEITABL3 127
Figure 65 Works per year as registered in the database WEITABL2 129

List of tables

Table 1 Some properties related to switch curve radius and angle of two switch types [Source: SBB] 10
Table 2 Switch geometrical type abbreviations and amounts 11
Table 3 Number of components per S&C type 19
Table 4 Comparison of static loads between transport modes and their respective infrastructure 31
Table 5 Typical values for the coefficients α, β and γ as a function of the track failures [ORE D161 (1988)] 36
Table 6 Estimated life of rails in North American Track from Sawley (2001) 47
Table 7 Parameters taken into account for the analysis 48
Table 8 Maintenance actions on a switch or crossings 52
Table 9 Renewal actions on a switch or crossings 53
Table 10 Track category description [SBB] 56
Table 11 Parameters influencing wear and degradation of switches & crossings [Jovanović &Zwanenburg (2002) and Scheffers (2007)] 57
Table 12 Economic (E) maintenance and renewal interaction 60
Table 13 Structural (U) and stochastic (O) dependence S&C components 61
Table 14 Input for the methodology: maintenance 62
Table 15 Input for the methodology: renewals 62
Table 16 Average replacement age and standard deviations for different S&C types (1988-2005)[DfA, SBB]... 77
Table 17 Number of replaced switches & crossings per type, of which the life in years and tonnage is known [DfA, SBB] 78
Table 18 Number of replaced switches & crossings per track category, of which the life in years and tonnage is known [DfA, SBB] 79
Table 19 Number of replaced standard turnout per track category [DfA, SBB] 79
Table 20 Combination of cumulative tonnage when replaced and amount of years in place [DfA, SBB] 81
Table 21 Correlation of parameters influencing standard turnout renewals 88
Table 22 Multi-parameter regression analysis (values to be multiplied by 10^7 to retrieve tonnage equivalents)..89
Table 23 Number of replaced diamond crossings with double slips per track category [DfA, SBB] 89
Table 24 Multi-parameter regression analysis (values to be multiplied by 10^7 to retrieve tonnage equivalents)..92
Table 25 Result of the parametrical test on complete switches and crossings 93
Table 26 Results for the modelling 95

Abbreviations

This list contains the abbreviations as used in this thesis. If the abbreviation originates from another language, an English translation is provided.

AW	*Ausweichung* S&C-type where small gauge track enters or leaves the broad gauge track (disregarded in this study)
DW	*Doppelweiche* S&C-type: Combined turnout
DKW	*Doppelte Kreuzungsweiche* S&C-type: diamond crossing with double slips
EPFL	*École Polytechnique Fédérale de Lausanne* (Swiss Federal Institute for Technology Lausanne)
EKW	*Einfache Kreuzungsweiche* S&C-type: diamond crossing with single slip
EW	*Einheitsweiche* Standard turnout
GD	*Gleisdurchschneidung* S&C-type: diamond crossing
HG	*Hauptgleise* Main tracks
LCC	Life Cycle Cost
LITEP	Laboratory for Intermodality, Transport and Planning
MS	*Mittelstück zu gekreuzter Gleisverbindung* S&C-type crossing part of a double intersecting crossover (disregarded in this study)
M&R	Maintenance and Renewal
NG	*Nebengleise* Secondary tracks
ORE	*Office de Recherches et d'Essais de l'Union Internationale des Chemins de fer(UIC)*
R135	Expression for break percentage of a train, here indicating 135%
RAMS	Reliability, Availability, Maintainability, Safety
S&C	Switches and Crossings
SNCF	*Société Nationale des Chemins de fer Français* French State Railway Company
SW	*Symmetrische Weiche* Symmetrical turnout
TSI	European Union's Technical Specifications for Interoperability
UIC	*Union Internationale des Chemins de fer*
ZV	*Zungenvorrichtung* Switch rail as used as a protection device to avoid wagons rolling unattended to get onto the main track. Passage in the false direction will generally lead to a dead-end track or immediate derailment (disregarded in this study)

1. Introduction, objectives and limitations

This chapter forms an introduction on this thesis. It includes an explanation of switches and crossings (S&C) and why they are needed, the objectives and the limitations of this study. Within this context a document layout is provided by means of an indication of the contents for every chapter.

1.1 The necessity of switches & crossings

Railway switches and crossings (S&C) are the devices which allow trains (cf. Figure 1):
1. to choose tracks to continue their way in different directions or v.v.;
2. to join multiple tracks or to split up a single track;
3. to change tracks to continue their way in the same direction but on different tracks;
4. to cross other tracks.

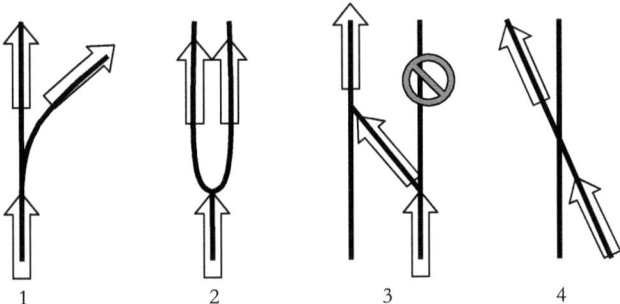

Figure 1 Generalized situation of the use of switches and crossings

Since the railway is a guided way (i.e. the track determines where the train will go, opposite to for example an automobile or a pedestrian) with only one degree of liberty (i.e. forward or backwards), S&C are crucial to uphold the geographical advantage of modern railways: the relative and absolute small amount of space needed -compared to motorways or waterways- to transport the same amount of passengers or freight. This efficiency can only be achieved by concentrating trains from different origins with the same destination at a certain point, or vice versa. It thus also reduces the actual track length needed, because trains can be concentrated on a smaller amount of tracks. In principle, what contributes most to this guiding principle is the wheel shape, with its inclined rolling surface for self-steering purposes and the wheel flange to avoid derailment and steer the wheel in the curves.

Another important contribution of S&C to the advantage of modern railways is the possibility they introduce, to divert trains to other tracks if the track which they were meant to use is blocked, for whatever reason (Figure 1, 3rd drawing). In this case the timetable can be upheld and the passengers or freight can still arrive at their destination, even though the original track is not available.

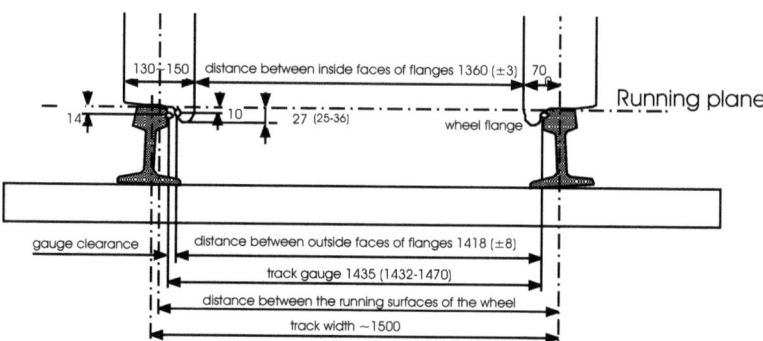

Figure 2 A train wheelset and the different track gauges and clearances

Although there exist various discontinuous possibilities to change or cross tracks, e.g. a turntable (Figure 3) or a travelling platform (Figure 4), in this study only continuous S&C are taken into account, for which counts that they form a continuous guided way which can be used by trains in all directions, without big efforts from or on the train or the infrastructure.

Figure 3 Example of a discontinuous way of changing tracks: a turntable in Muttenz (CH)

This also means that discontinuous S&C which cause a derailment if used in the "wrong" directions (the direction for which the point rails are not set) are also not taken into account. Annex 1 to this report shows such a switch.

Also not taken into account in this study are adjustment or expansion switches (Figure 5).

A **detailed description of the S&C** that are taken into account, including a description of their components can be found in **chapter 2**.

Figure 4 Example of a discontinuous way of changing tracks: a travelling platform in Rotterdam (NL) [1]

Figure 5 An expansion switch, Lyon (F)

1.2 The reasons for this study

As indicated above, switches and crossings (S&C) have an important function within the railway network. They are an integrated part of the track part of the infrastructure together with straight and curved tracks, the last ones in this thesis referred to as "plain track". Other parts of the infrastructure can be embankments, tunnels, bridges, drainage, signalling and automatic train protection equipment, over-head lines and other components needed for electrification, road-crossings, noise barriers etc.

All these components are subjected to **wear, tear, plastic deformation and/or degradation**, phenomena described in **chapter 3**. That might be because they are subjected to a certain load, or due to weather influences. This phenomenon has a functional limit: at a certain moment a component cannot wear, tear or deteriorate any further because of availability, safety, comfort and/or economic reasons. Not a single component therefore has an ever-lasting life. This results in a continuous **maintenance and renewal** demand (described in **chapter 4**) for the infrastructure of a heterogeneous railway network.

---

[1] © portpictures.nl, published with permission by e-mail d.d. 27 May 2008

The annual amount of money used for the maintenance and renewal of all railway tracks is significant – at the Swiss Federal Railways (SBB) multiple billion Swiss Francs per year[2]. The proper attribution of this budget is a challenging task for the infrastructure manager when taking into account the properties influencing the decisions, a non-exhaustive list of these might be:
- Morphological properties of the network:
    - the size of the network
    - the structure of the network
- Working constraints:
    - the availability of properly trained personnel
    - the availability of the machines to carry out the work
    - the often difficult access to these sites
    - on saturated or close-to-saturation networks: the availability of time to carry out maintenance
- Financial constraints

Some of the arguments above are related to a unique feature in the railways: when carrying out certain maintenance or renewal works on the tracks, those tracks are not available for train services. In other words: when replacing a rail, trains simply cannot run on that specific track. Rerouting is in some way or the other possible, but with much more difficulty in time and place than e.g. on a motorway. A train cannot be rerouted on a parallel track if the signalling system is not suited for trains in 2 directions, neither is it available if it is used by a train in the opposite direction.

This rather straightforward description of practical constraints can be summarized in two major strategical achievements for the SBB infrastructure management, as presented by Putallaz (2007):
1. A sustainable management of the railway infrastructure.
2. To maintain a saturated railway network.

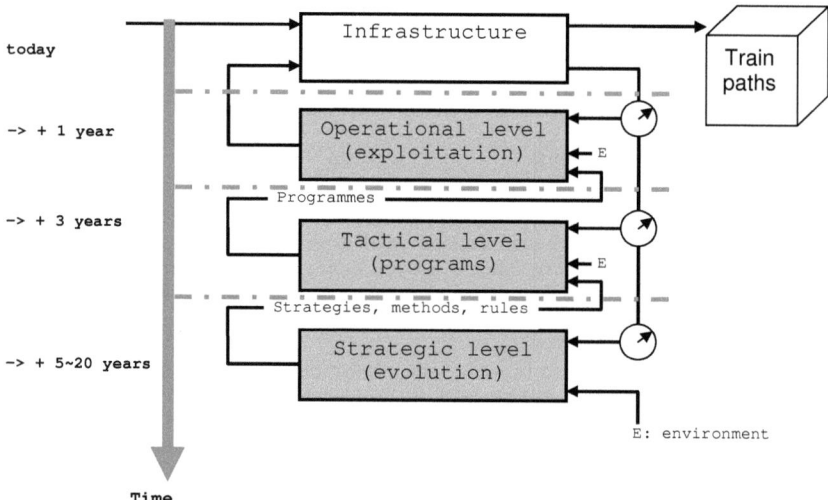

Figure 6 Organisational model, Putallaz (2007)

---

[2] Note that there are several infrastructure managers in Switzerland, e.g. the Lötschberg railway line is managed by the BLS

With these two points in mind, including the abovementioned non-extensive list of constraints, proper maintenance and renewal budget attribution has not been an easy task. To help the infrastructure manager, several maintenance management tools have been developed, e.g. Decision Support Systems like RailSys or EcoTrack, as described in Rivier (1992) and Ferreira & Murray (1997). These can help at all layers of the organisation: operational level, tactical level as well as strategical level (Figure 6).

The infrastructure manager is exposed to a high amount of information about the state of its network. For different organisational levels (operational, tactical, strategical), different kind of information is necessary to plan:
- the strategies, methods and rules to initiate evolution at the strategical level;
- the from this evolution resulting programs to at the tactical levels;
- the operational planning in the operational phase, mainly focussed on execution of the works with respect to the train exploitation.

The different levels are exposed to the environment, influencing them and providing opportunities and restrictions. The final products are train paths: a timetable in which time and space are reserved for a certain train, which the infrastructure manager can sell to the railway company running the trains. It has to be taken into account that inspection, maintenance or renewal works during which the tracks are not available can also be regarded as train paths not being available for sail, thus representing a certain negative value for the infrastructure management.

The result of the use of decision support models in infrastructure management is visible as a large rationalisation of information collection and processing (e.g. site visits, use of the measuring trains) and the maintenance and renewal planning, on networks where these tools are in use, during the last 10-20 years. This, however, only counts for plain track, i.e. straights and curves, and tunnels, bridges, signalling and overhead electrical equipment, not for S&C.

Different reasons for this lack are identified in Jovanović & Zwanenburg (2002). The main reason is that, while for plain track, for different geometrical degradation modes (e.g. cant, twist, level, gauge) and component deterioration (e.g. cracks, gauge corner cracking, head checks, corrugation) the relation with track parameters (e.g. sub-soil quality, type of sleepers) and train load (e.g. number and amount of trains, axle loads) is known, as proven by Veit (2005), the knowledge on degradation and deterioration processes of S&C has remained rudimentary. It is the knowledge on these degradation and deterioration processes which is necessary for the development of the maintenance management tools or for maintenance optimisation in general as described by Dekker (1995).

The current way of collecting information on the state and wear of S&C is through railway employees on foot, who visit a switch or crossing and then check if the degradation and wear parameters are within their limits. Mainly due to this, the exact information on S&C degradation cannot be retrieved at this moment, although developments have been commenced to rationalise and digitize this type of data collection. What remains possible, however, is providing the relationships between the expected lifetimes of S&C and their components, and parameters which effect this lifetime. The Swiss Federal Railways (SBB) –to which this study is limited, but in case appropriate also knowledge from abroad is taken into account– have databases on S&C and S&C-components maintenance and renewal works which are complete since 1997. This information can be coupled with other databases, describing the use of the tracks (e.g. track loads (cumulative tonnages) since 1950, train speed, type of trains) and other influencing parameters (e.g. soil quality).

The reason for this study can thus be summarized as:
The identified need to develop degradation and wear models for railway switches and crossings taking into account and quantifying influencing parameters, to optimise the forecast of S&C maintenance & renewal needs.

This thesis will present these models. The **methodology to derive these models** is described in **chapter 5** and the **models themselves** in **chapter 6**. The **model testing** is presented in **chapter 7**.

The optimisation of the forecasting of the S&C maintenance and renewal needs can be carried out at different levels:
- Component level:
  o foreseeing when e.g. sleepers, rail or other parts need replacement;
- Complete switch or crossing level:
  o foreseeing when a complete switch or crossing needs replacement, based on the fact that at a certain moment different components are in need for replacement;
- Local level:
  o foreseeing when e.g. several switches and crossings or similar (or not) components of different switches and crossings in a certain station area need replacement, the works can be combined to reduce costs;
- National level:
  o foreseeing peaks in maintenance and renewal needs on the network for the mid term (ten years) on which can be anticipated by spreading the work over more years;
  o strategic decision making on which S&C perform better than others and coupling that back towards the S&C investment strategy.

Improvements can be researched at operational, tactical as well as strategical level. As far as maintenance and renewal is concerned, these three are related to time scales of respectively 0-3 years, 3-10 years and 10-25 years or longer. For different reasons already beforehand a tactical optimisation is chosen:
- optimisation at purely operational level is more related to improving e.g. work methods, logistics at the site and materials used – another research topic, but neglected here;
- the fact that data is available for only the last ten years and thus the reliability of forecasting maintenance and renewal needs might reduce for periods longer than ten years when purely focussing on strategical level.

Although Jovanović & Zwanenburg (2002) concluded that a large part of the S&C malfunctioning are due to failures of the actuator or the locking of the switch blades, this thesis only focuses on the mechanical wear of the components supporting and guiding the trains. This is also due to the fact that these last components are the responsibility of the permanent way department, and the switch engines often from the signalling department. For the same reason also the switch heating is not part of this thesis. This can also be regarded as more daily operations related topic – since failures to these devices can nowadays be registered online via condition monitoring systems.

1.3 Derived objectives of this study

With the above mentioned models the main objective is to **determine the S&C maintenance and renewal needs for the next ten years** (mid-term planning). The resulting costs of S&C maintenance and renewal for different load scenarios (faster trains/slower trains, more trains/less trains, increase in axle load) can be determined. However other analyses are also possible when a cost-database is available, e.g. the following LCC comparisons:
- between similar switches or crossings subjected to different loads,
- between different switch or crossing types subjected to the same load,
- optimisation of combined S&C lay out to reduce long term costs.

1.4 Terminology

In this document the expression "switches & crossings" with its abbreviation S&C[3], will be often used. In principle, this is the American term for the devices which forms the main subject of this thesis. The "British English" term for a switch is "set of points" or "points". A synonym for switch, but of which the origin is less clear, is "turnout".

To make it even more difficult, a "crossing" can also be a part of switch or crossing, for which in the United States the word "frog" (cf. Figure 20 and §2.4.2) is used. In this thesis the word "frog" is used for this component to distinct the two.

The term "degradation" is used for a decrease in track geometry quality and the term "wear", "tear" and "plastic deformation" for a reduction in the quality of track components. In the last case, often only "wear" is written where wear, tear *and* plastic deformation are mentioned. If it is different it is mentioned separately.

---

[3] A list of abbreviations is provided at the beginning of this thesis

2. Description of switches & crossings

This chapter provides a state of the art in form of a description of the different types and categories of switches and crossings (S&C), their use and their respective components.

2.1 Introduction

The fact that national railways have had the liberty for years to design and sometimes even manufacture their own Switches and Crossings (S&C) has led to a high amount of different designs. Therefore, composing a catalogue of all existing switches and crossings is not an easy task. BWG Butzbacher Weichenwerke, a known German S&C manufacturer, now part of the Austrian VAE steel group, even stated, when demanded, that they did not produce a catalogue, because the demands are so specific for every order and (national) railway system that standardisation is hard to achieve and buying of-the-shelve still an exception.[4]

Though, still, two categorisation types are used: by geometrical properties, or by component type.

The geometrical properties can be
- curve radius in diverting direction:
  o e.g. 160m, 185m, 200m, 300m, 500m, 900m, 1600/2600m[5], 10000/4000m, 12000/6100m;
- diamond crossing angle (expressed in m/m):
  o e.g. 1:7, 1:8, 1:9, 1:10, 1:11, 1:12, 1:14, 1:16, 1:19, 1:21.5, 1:24, 1:25, 1:32.05, 1:38, 1:42, 1:6/1:9[6];
- type:
  o e.g. standard turnout, symmetrical turnout, combined turnouts, diamond crossing with or without single or double slips.

Based on the type of components used, the following distinctions are made:
- by sub-base material
  o ballasted track
  o on a concrete slab track
    ▪ on soil
    ▪ on a bridge or in a tunnel on the tunnel floor
- by sleeper type
  o steel
  o wood
    ▪ treated soft or hardwood; now less common since the original treatment with a distillate of coal tar causes the pollution of soil and air with extreme carcinogen substances. The substance is now replaced by more environmental friendly, but also more expensive substances;
    ▪ tropical hardwood; although suspect of being not environmental friendly, in exceptional cases still used for less standard or more complicated S&C. Also used for S&C positioned above tunnels, on viaducts or in the neighbourhood of other structures, where a limited ballast height is available;
  o polymer compositions indicated as "artificial wood"; a new development combining the properties of wood, with the durable properties of polymers as

---

[4] Mrs. S. Szymanski of BWG per e-mail, April 2006 and confirmed by Mr. Höhne of BWG, October 2006
[5] Double curve radii are common in large turnouts
[6] Different switch angles are possible in curved diamond crossings

described in Koller (2008), it is not used in Switzerland and therefore ignored in the rest of this report;
- o concrete; most commonly used nowadays for new, standard S&C in Switzerland and other countries;
- by rail type: normally expressed by an expression which includes the weight in kg/m, e.g. 60E1 for a 60 kg/m rail of type E1 (also E2 exists with a different gauge corner).

The categorisation and the different components of switches and crossings are described in detail in the next paragraphs.

2.2 Categorisation by means of curve radius or switch angle

The main indicator for the way in which a switch can be used is the switch radius, or –related to the radius– the frog/crossing angle. This has to do with the curve radius used in the curved direction: this radius is the parameter which determines the maximum speed with which the trains can roll in the curved direction. Often in switches and crossings the most small curve radii of the railway networks can be found: down to only 185 meter in the most commonly used standard turnout on Switzerland's main line tracks. To compare: the minimum radius in plain track curves on the Gotthardline is 300 meters and 500 meters on the Lötschberg line (over the pass and through the old tunnel, not the base-tunnel line opened in 2007). This small curve radius causes transversal accelerations (non-compensated lateral accelerations) a_{nc} on the moving trains in the curved direction. For reasons of comfort and safety, these accelerations, which are also exceeded on the passengers, cannot be too high and therefore the speed of the train in the curved direction has to be reduced.

In curved plain tracks, normally *superelevation* –also known as *cant*– is applied: the outer rail is positioned higher than the inner rail to compensate (part of) the transversal accelerations to which the train and it passengers or cargo is exposed. In switches and crossings this is not applied; it is almost impossible to introduce if not yet the track before and after the switch or crossing are already superelevated. This therefore imposes an extra speed restriction compared to plain track curves.

Finally there is a third interest peculiarity at switches and crossings and especially those with a small radius: the transition curves at which the curve radius gradually reduces from infinite (i.e. a straight piece of track) to the small-radius-curve is extremely short. This causes high changes in transversal accelerations, known as the *jerk* Ψ and expressed in m/s³. These even have a higher impact on comfort reduction as the transversal accelerations themselves.

Table 1 Some properties related to switch curve radius and angle of two switch types [Source: SBB]

Type (cf. §2.3)	Approximate. switch length (m)	Curve radius (meter)	Available switch angles	Max. speed in curved direction passenger[7]/freight[8] trains (km/h)	jerk Ψ (m/s³) [9]	Transversal accelerations a_{nc} (m/s²) [10]
Standard turnout	26	185	1:7, 1:8, 1:9	40 / 40	0.39	0.67
Standard turnout	33	300	1:9, 1:12	55 / 50	0.63	0.78
Standard turnout	42	500	1:12, 1:14	65 / 60	0.62	0.65
Standard turnout	56	900	1:16, 1:19	95 / 90	1.08	0.77
Standard turnout	74	1 600	1:21.5	115 / 110	1.08	0.64
Standard turnout	83	1 600/2 600	1:24	125 / 120	0.85	[11] 0.75
Symmetrical turnout	22	200	1:7	40 / 40	0.36	0.62

[7] Swiss standard train categorie R
[8] Swiss standard train categorie A
[9] Calculated for a 18.7 meter long SBB wagon of type EW1/VU1, maximum permitted value in Switzerland 1.2 m/s³; in Austria 1.0 m/s³
[10] Maximum permitted value in Switzerland 0.8 m/s²; in Germany and Austria 0.65 m/s²
[11] For the part with 1600 meter radius. For the other part with larger radius, a smaller value is obtained.

Related to the switch crossing curve radius is the switch or frog angle. It is an expression for the angle of the crossing rails at the frog (cf. §2.4.2). Larger angles will be found at switches and crossings with smaller curve radii. This relationship is explained in Table 1 for some common switches. This table also includes the length of the switch, the maximum calculated jerk and maximum designed transversal accelerations for these switches. The jerk depends strongly on the rolling stock used. Shorter rolling stock will experience a higher jerk than longer rolling stock. The relatively old rolling stock for which the jerk is calculated here is nowadays largely replaced by longer –thus more comfortable, regarding the jerk in a switch– rolling stock.

As can be seen in the table above, a length of 83 meters is necessary for a switch which allows speeds in the curved direction of more than 100 km/h. It is not difficult to imagine with this number in mind, that switches and crossings allowing speeds of 200 km/h or higher in the diverting direction have lengths of over 200 meter.

Curve radii smaller than 250 meter make track gauge widening necessary to guarantee a smooth running of the train wheels through the track.. This is thus applied in switches and crossing where this is the case.

2.3 Categorisation by means of type

A distinction can be made by S&C type as presented in the figure below, i.e. by their function and their geometric properties.

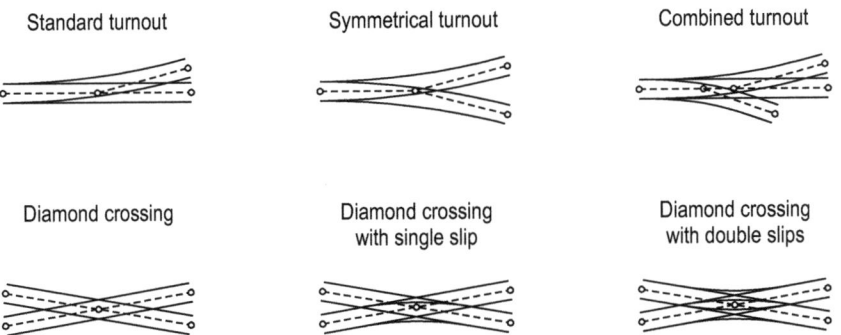

Figure 7 Categorisation of S&C types by their function

The amount of these switches and crossings in the database are described in Table 2.

Table 2 Switch geometrical type abbreviations and amounts

	Abbreviation used in this report[12]	Number in database
Standard turnout	EW	12 874
Symmetrical turnout	SW	342
Combined turnouts	DW	367
Diamond crossing	GD	107
Diamond crossing with single slip	EKW	64
Diamond crossing with double slips	DKW	1 308

---

[12] The abbreviations are based on the german terms for the respective switches and crossings

Some of the switches and crossings of which data is available in the database are discarded here:
- all connections and crossings with narrow gauge tracks (5 in the database, too small sample size and special constructions);
- protection switches, which are installed to avoid runaway wagons or SPAD[13]s to enter the main tracks. The curved direction is normally not used, except for the run away-wagon or train which passed a red signal (20 in the database).

The properties of the different geometrical types are described below.

2.3.1 Standard turnout

A standard turnout is the most common used switch in railway networks. It allows the train to be directed in different directions, or to combine two tracks in to one. A similar functionality of this type can be found with the symmetrical turnouts (cf. §2.3.2) and the curved turnouts (cf. §2.3.8)

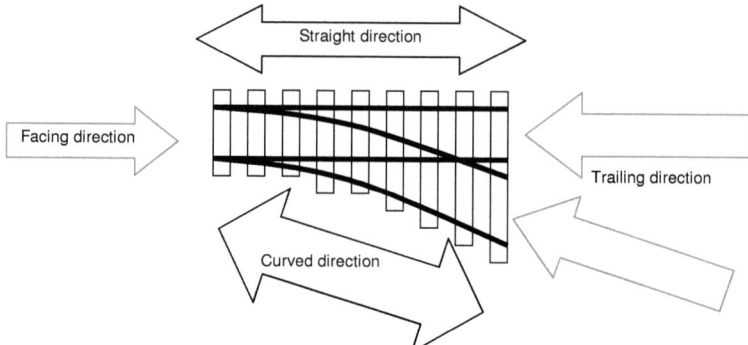

Figure 8 Directions at a standard right-hand turnout

Of a standard turnout one direction is straight (also known as the "through"-direction) and the other direction is curved (also known as the "diverting"-direction). It is called a left hand turnout when the curved direction is to the left (cf. the standard turnout in Figure 7) and a right-hand turnout if the curved direction is to the right (cf. Figure 8), in the direction facing the moveable point rails. These moveable points allow the vehicle wheels to be continuously guided in one of two directions. They are set with a switch actuator, which can be an electrical motor or a pneumatic or hydraulic driven device. The actuator or motor is not part of this study.

---

[13] Signals passed at danger: term for red signal passage of a train

Figure 9 Switch actuator for moving the switch blades, Rotterdam (NL)

Neither part of this thesis are the switch heating devices, which heat electrically, with gas burners or with a central heating system (including salt water through pipe-lines) the moveable point rails during cold days, to avoid the moveable switch blade freezing on the stock rail, the seats or the rollers on which they are posed.

In the straight direction, the maximum permissible speed is the normal running speed as on the plain track before and after the switch - speed records with trains have confirmed, that much higher speeds than used during normal exploitation can be safely carried out in the straight direction on existing high speed switches.

For the diverting direction the story is different. Here the maximum allowed speed depends on the curvature or switch angle (cf. §2.2). A moving train in this curve is exposed to transversal accelerations which results in a reduction in the passenger comfort or instability for the freight in or on the train. There is a maximum prescribed for this acceleration and this thus determines the maximum speed allowed in diverting/curved direction of a switch.

When deciding for the installation or replacement of a standard turnout, it is first of all the available space which determines which new turnout type can be used. If there is no restriction regarding the available space, it is the necessary speed in the diverting/curved direction, from an exploitation point of view that can be decisive. A higher speed in a curve generates higher transversal accelerations and thus demands a bigger curve radius and smaller angle to reduce this. From a technical point of view, trains, normal railway tracks and S&C can withstand higher transversal accelerations than currently allowed in railways, but only in combination with a significant reduction of the passenger comfort.

2.3.2 Symmetrical turnout

The symmetrical turnout is a special standard turnout, where both directions are curved in opposite direction with the same radius. A reason for using this type rather than a standard turnout might be that the switch is already situated in a curve and the other direction curves into the opposite direction, with a limited amount of space around the track. For the counter curve this introduces a serious speed restriction, since immediate succession of the counter curve after the curve will result in a 180 degree change in direction of the transversal accelerations. It also has geometrical limitations.

Another reason might be that two parallel tracks are combined to one track, the latter situated exactly in the middle of the prolonged centre lines of the two parallel tracks. Similar as written above, also here the resulting S-curves to get both directions together at the symmetrical turnout, introduce a speed limit for all directions, because the change in transversal acceleration (from left-curve to right curve or v.v.) would otherwise be very uncomfortable

A third and final reason might be that it limits the amount of space needed, when different switches and crossings are placed right after each other, and on which the trains using this switches have to remain at their same speed in different directions. This is the case at a marshalling yard.

Figure 10 Symmetrical turnouts, marshalling yard of Villeneuve (F)

The wagons rolling from the shunting gradient due to gravity, should be maintaining the same speed in all directions. The curved direction on a turnout causes more rolling resistance than the straight direction, which results in different rolling characteristics for the different directions: a situation to be avoided on marshalling yards in the past. Nowadays this effect has been reduced, due to modern systems able to measure the speed of the wagons in real-time and regulating the trackside braking-system towards this problem.

2.3.3 Combined turnout

When the available space is limited but multiple tracks have to be deviated from the main track, a combined turnout might be the solution. A combined turnout is a multiple standard turnout intertwined and in case it are two standard turnouts combined, it has two sets of two points. The points of the second turnout are placed before the frog/crossing of the first one. The complexity increases not only because -in the case of a double standard turnout- of the four different directions that are combined, but also because of the extra frog necessary in the switch. Use of combined switches in main tracks is therefore limited and they are only found at:
- station entrances (mainly because platform extensions reduced the amount of space available for standard turnouts);
- marshalling yards where sometimes also limited space is available but still track lengths should be as large as possible
- ferry boat quays where sometimes 5 tracks have to be deviated from one on an extremely short track to connect to the 5 tracks on the ship itself.

Figure 11 Combined turnout, marshalling yard of Sibelin (F)

2.3.4 Diamond crossing

The diamond crossing allows two tracks to cross each other. The standard type has no moveable parts, since there is no need to choose a direction. The fact however that two times two rails have to cross each other makes 4 places necessary where gaps in the rolling surface have to be introduced to let the wheel flange pass (cf. §2.4.2 and 2.4.3 for an explanation): two frogs and two obtuse crossings.

Because of these obtuse crossings, which form large gaps in both running rails, the use of all diamond crossings and derived crossings (i.e. diamond crossing with single slip, diamond crossing with double slips), is limited to speeds up to 140 km/h in the straight direction in Switzerland.[14]

Figure 12 Diamond crossing in the junction of St Benoit (F)

2.3.5 Diamond crossing with single slip

This is a combination of a diamond crossing and two standard turnouts. It means that two of the four ends of a diamond crossing are connected by a curved section as with a standard turnout. The diamond crossing with single slip is therefore equipped with two sets of moveable points.

---

[14] Other values might be found for other countries

Figure 13 Diamond crossing with single slip, Zwolle (NL)[15]

2.3.6 Diamond crossing with double slips

This is a combination of a diamond crossing with four turnouts. It allows trains from all of the four ends to continue in both two opposite ends. To achieve this, the diamond crossing with double slips is equipped with four sets of points.

Figure 14 Diamond crossing with double slips, Rotterdam IJsselmonde (NL)[16]

2.3.7 Special combinations

<u>Cross-over</u>
To allow trains to change tracks when there are two parallel tracks, a cross-over is used. This is a combination of two standard turnouts. When changing from the right track to the left track, two left-hand turnouts are used ("left hand" means the diverting direction is to the left, as seen facing the points – explained in §2.3.1). When changing from the left track to the right track, two right-hand turnouts are used.

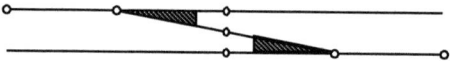

Figure 15 Schematic cross-over with two right-hand turnouts

These combinations of turnouts are preferred to have the same angle and/or curvature, since the speed is determined by this and the opposite transversal forces will then be the same for both curve and counter curve and the same in both directions.

<u>Double intersecting cross-over</u>
If both left-right and right-left connections of two parallel tracks are combined, a combination of four turnouts and a crossing is obtained. This reduces the amount of track length needed for S&C and is often used before and after platform tracks to allow all possibilities regarding the use of the platforms on a station (a train from any direction can use any platform track).

---

[15] © Arie van Zon, published with permission by e-mail d.d. 20 June 2008
[16] © Arie van Zon, published with permission by e-mail d.d. 20 June 2008

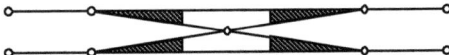

Figure 16 Schematic double intersecting cross-over with two right-hand, two left-hand turnouts and a crossing

Figure 17 Double intersecting cross-over at Rotterdam (NL)

Right-angle crossing
Situations exist, at which crossings are necessary which let tracks cross each other at a 90-degrees angle or an almost 90-degrees angle (oblique crossing). The right-angle crossing can only be passed at a low speed, since there is an opening in the running surface, which the train wheels have to "jump". Especially for right-angle crossings close to or at 90 degrees, this jump is at exactly the same time for the left and the right wheel, creating dynamic loads on both train and track.

Figure 18 Almost right-angle crossings at Utrecht Blauwkapel (NL)

2.3.8 Special geometrical property: curved switches

A standard turnout has one curved and one straight direction; a symmetrical turnout two oppositely curved directions. When applied in curves sometimes both directions will be curved in the same directions. Although some national railway networks have the "luxury" to be able to position switches and crossings always at straight stretches of track, allowing one direction of a turnout to be straight and one to be curved, typically in mountainous areas as Switzerland, switches or crossings simply have to be placed into curved sections of plain track. This can be also the case when station platform extensions occur: although 100 years ago, when in the initial design of a station the complete station and its entrances were positioned on a straight piece of track between two curved sections, platform extensions to serve longer trains (maximum international standard for passenger trains currently 400 meters) cause the station entrances, with all the switches and crossings to guide the trains to different platforms, to be pushed backwards into the curved plain track sections.

Figure 19 Left: Special curved diamond crossing with single slip in Bern (CH). Right: Curved standard left hand turnout in Palézieux (CH)

2.4 Components of switches & crossings

This paragraph explains the switch and crossing components more in detail. These are:
- the switch blades and the accompanying stock rail, including the sliding chairs
- the frog
- the check rail with the cross-bar
- the common crossing (only for diamond crossings, diamond crossings with single slips and diamond crossing with single slips)
- the intermediate rails
- the sleepers
- the ballast

For a part of these components, their exact position in a switch can be found in the figure below.

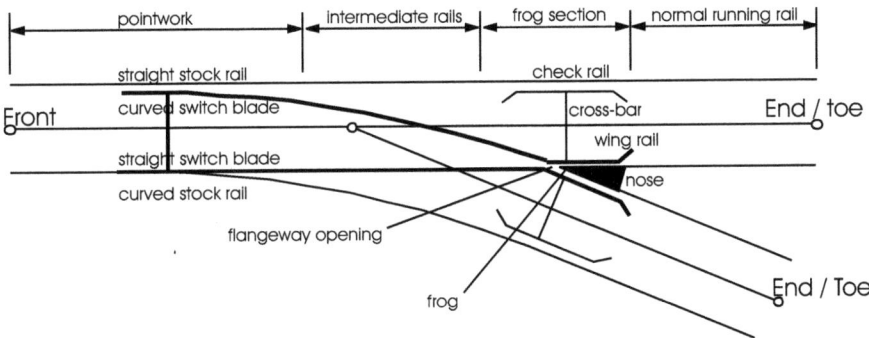

Figure 20 Standard right-hand turnout with indication of the important components

The exact number of these components also depends on the switch or crossing. The table below shows their presence in the different types of switches and crossings as presented in the previous paragraph.

Table 3 Number of components per S&C type[17]

Component	Standard turnout	Symmetrical turnout	Combined turnout (double)	Diamond crossing	Diamond crossing, single slip	Diamond crossing double slip
Switch blade + stock rail	2	2	4	-	4	8
Frog	1	1	3	2	2	2
Check rail	2	2	6	4	4	4
Common crossing	-	-	-	2	2	2
Intermediate rail between wing rail and switch blade - straight	1	-	*	-	-	-
Intermediate rail between wing rail and switch blade - curved	1	2	*	-	-	-
Switch actuators/motors	*1	1	2	-	2	2
Switch heating units	2	2	4	-	4	8

For this table, it is presumed that all S&C are posed on sleepers and a ballast bed or on a slab track.
The number of intermediate rails in a combined turnout depends on the way in which different turnouts are combined. The number of actuators on a standard turnout depends on the length of a turnout: longer switch blades are often powered by multiple actuators. Also a frog with a moveable nose (cf.§2.4.2) often has its own actuator/motor.

2.4.1 Switch blades and accompanying stock rail

The switch blade (a.k.a. point rails) and the accompanying stock rail are specially devised long metal rails, with a continuous different shaped area depending on where the two pieces are cut in half. This has to do with the fact that a train wheel will have to change from one rail to the other (or not). This is done by introducing a new rail –the switch blade- between the stock rail and the

---

[17] the value * means "depends on the configuration"

wheel flange, and then gradually taking over the function of supporting the train wheel from one rail by the other (Figure 21)

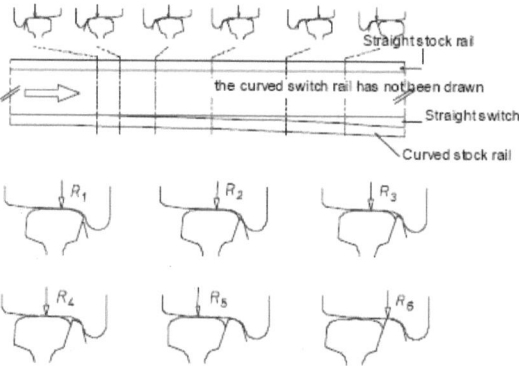

Figure 21 Schematic view of the support of the wheel by the stock rail and in R6 the switch blade

Switch blades were previously devised from standard rail profiles, but nowadays special rail profiles are used from which, in an easier way, a switch rail can be produced by ways of grinding of metal layer-by-layer. The pointed head as visible at R1 in Figure 21 is gradually widened towards a normal shaped head introducing itself in the area between wheel flange and the vertical running surface of the stock rail.

The change of support from R5 to R6 in Figure 21 causes dynamic loads in the train wheel and thus in both train and track. This is a cause for discomfort, noise and wear and thus a reason for costs. One of the ways to reduce this dynamic load is gauge widening in the region of the switch blades, which has as a positive side effect that it reduces at the same time that the wheels running against the switch blades (another cause of wear). One of those systems is called FAKOP, an abbreviation for the German *Fahrkinematische Optimierung* [running kinematics optimisation].

Two types of switch rail exist: high profile and low profile. The low profile switch rails have the advantage that there is enough space for the rail seats, sliding chairs and/or rollers -necessary for the movement of the blade- and a higher vertical stability. However, between the low-profile blade and the high-profile intermediate rail, a special construction is necessary to adapt the rail height (Figure 22).

Figure 22 Change from high-rail profile to low rail profile

The switch blades are attached to the intermediate rails by means of:
- electric or thermo-welded welding;
- a normal joint and fishplates or;
- an insulated joint.

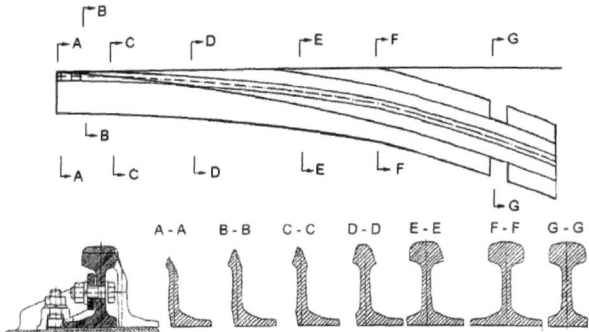

Figure 23 Different cross sections of a high-profile switch blade

The length of the switch blades depends on the switch angle. Smaller angle switches and crossings with larger curve radii have longer switch blades. On switches and crossings for high speed lines with speeds op to 200 km/h in the diverting direction, the switch blade length is approximately 60 meters.

The movement of the switch blades –necessary to allow the trains to be directed in different directions– is executed by one or multiple (in case of long switch blades) switch actuators, which are connected to the blades by driving rods. To support the rail during the movement, sliding chairs are placed on top of the sleepers. Older types of switches and crossings are equipped with metal sliding plates which need regular greasing. Newer types do not need greasing anymore. Another reduction in the greasing demand has been achieved by the introduction of rollers which also make the movement of the switch blades easier.

The movement from itself is reached by bending the switch blade, i.e. there is no rotation point, but rather a certain length over which the blade is bent. To facilitate this bending, sometimes the rail foot width is being reduced over a certain length (cf. cross section G-G in Figure 23).

2.4.2 Frog

At the frog[18] –also known as the crossing part or switch diamond– two rails cross each other. To allow the passage of the wheel flange (which is lower than the running surface of the wheel, cf. §1.1), a gap in the running rail has to be introduced: the flange groove. This introduces a discontinuity in the running surface (around b in the figure below), which is overcome, by letting the running surface "drop" from the wing rail to the nose. A possible derailment at the height of

---

[18] This name is probably related to the fact that this component has four rails attached to it, similar to the two arms plus two legs of a frog.

the gap is avoided by applying check rails, which reduces the play transversal to the rail of the wheel on the opposite site of the frog. Therefore every frog is equipped with two check rails, one for each direction, sometimes connected with the frog with a cross-bar. The check rail is explained more in detail in §2.4.3.

A frog can be made in different ways. The three standard ways are:
1. constructed from special rail profiles, which are ground in their appropriate form and then bolted together;
2. constructed from special rail profiles, but with a casted manganese nose;
3. a completely casted, manganese frog in one piece.

Figure 24 Left: constructed frog type 1 in Rotterdam (NL); right: cast manganese frog in France

SBB normally uses type 1 and 2, but other countries prefer the third type. The use of the cast manganese has to do with its built-in resistance to the hammering effect of the train wheels, when they jump from the wing rail to the nose of the diamond crossing: this causes the manganese to get even tougher, better resisting future loads.

The accompanying disadvantage is that a fracture in the cast manganese frog is difficult to repair by welding and makes a complete replacement often necessary. It is for this reason that often still for a constructed frog (type 1 and 2) is chosen, because of the possibility to replace it in even smaller components. Another disadvantage of the cast manganese frog is also that a new mould is necessary for all different switch geometries. A constructed frog can simply be made with the same tools.

Normally, the two crossing rails at a frog cross each other straight. However, when the curve continuous in the same direction after the switch (e.g. Figure 1, drawing 1), it might be convenient to apply a curved frog.

The frog is attached to the four rails by means of electric or thermit weldings or an (insulated) joint.

Swing nose crossing/moveable frog
As expressed above, a special property of a frog is the fact that the wheel jumps over the flangeway gap from the nose to the rail or v.v. At this point the wheel remains more or less supported vertically: the support is gradually taken over from the nose by the rail, but these points are not on the same part of the wheelband, still resulting in a jump. However, transversal there remains the flangeway gap of which derailment is only protected by means of the check rail opposite of the frog, guiding the complete wheelset. It is not difficult to understand that with axle loads of over 22 tons, the jump is a cause for dynamic forces and the derailment risk provokes strict maintenance and renewal requirement in case o wear and plastic deformation. With the dynamic forces (they are like hammering) also comes an amount of noise and vibration.

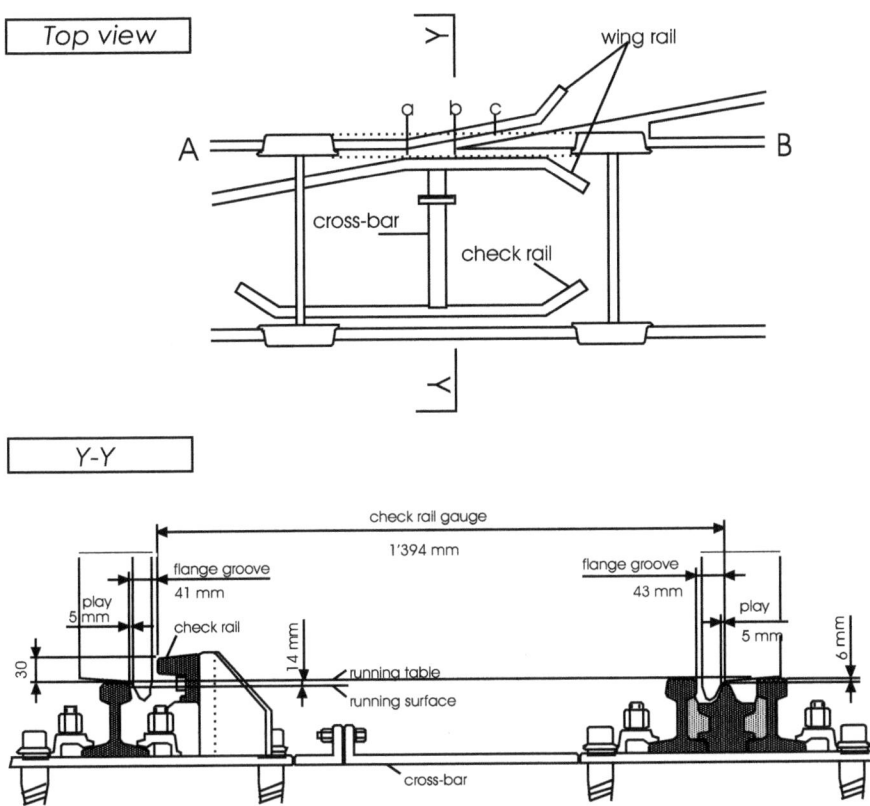

Figure 25 Schematic view of the passage of a wheel through the frog including clearances and gauges

(a = wheel supported by rail, b=jump of wheel to the nose, c=wheel supported by the nose)

Beside this, switches and crossings with larger radii to allow higher speeds in the diverting directions, thus with smaller crossings angles (cf.2.2) made the flangeway opening continuously longer. The combination of all these reasons has led to the development of crossings with a moveable frog or, also known as swing-nose crossings. These devices move the nose against the wing rail, to close the flangeway gap. It thus supplies a continuous running surface and continuous guidance. Due to this, the wheelband will always be in contact with the rail on the same spot. The result is a reduction of the dynamic load and a more smooth passage of the wheels through the frog part.

The movement is achieved by equipping the swing nose with a separate engine, or by mechanically connecting the nose by means of rods, to the same actuator as for the switch blades.

Figure 26 Left: Swing nose crossing, Amsterdam (NL)[19] and Wanzwil (CH)[20]

One of the first places where the swing nose crossings could be found was on high-speed lines, where the flange way opening induced an extra derailment risk which was to be abandoned. In general on all tracks for speeds higher than 160 km/h (France: 220 km/h) the switches are of the swing nose-type. The only exception is the upgraded line between Berlin and Hamburg in Germany, where switches with normal frogs are installed in tracks which allow 230 km/h.

Another use of the swing nose is in urban areas where the reduction in dynamic loads results in less noise and vibrations. For example the docklands light railways has a lot of switches with a swing nose for this reason and for reasons of reliability for the automatically driven trains.

In principle, a frog with a swing nose, could do without the check rail (cf. §2.4.3 since there is no flangeway opening. But it remains an exception, when the check rail is not placed.

2.4.3 Check rail

The check rail forms part of all switches which have openings in their running surface. They are placed opposite of these openings to avoid a wheelset taking the wrong route in this flangeway opening. They gradually reduce the free space a wheel flange has for transversal movement, to a minimum exactly opposite of the opening.

Figure 27 Check rail on the left, opposite of the frog on the right, Rotterdam (NL)

---

[19] © Arie van Zon, published with permission by e-mail d.d. 20 June 2008
[20] Source: SBB Regulation R I-22220

The functioning of the check rail can be seen in Figure 25. The top of the check rail is in almost all countries a couple of centimetres higher than the running surface of the rail. Only in the Netherlands the top of the check rail is at the same height as the running rail.

2.4.4 Common crossing

The common crossing has a lot of similarities with the frog (cf.§ 2.4.2): it allows two rails to cross and therefore has openings in the running surface to allow the wheel flange to pass. Also the 3 ways in which they can be composed are the same: constructed, constructed with cast manganese parts or casted manganese.

Figure 28 Cast manganese common crossing, Junction of St. Benoit (F)

The difference with the frog, which is used in all types of turnouts and crossings is that it only exists in diamond crossings and its derivatives: the diamond crossings with single slip and the diamond crossing with double slips. Most particular property is that in the case of a common crossing, exactly opposite, on the other running rail, there is another common crossing. This results in the common crossing always being equipped with a high metal plate on the inside, as a replacement for the check rail which normally avoids derailment at flangeway openings.

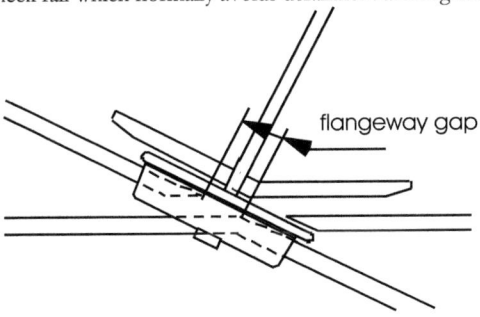

Figure 29 Schematic representation of a wheel passing a common crossing

Special type: diamond crossing with moveable switch blades in the common crossing
As stated in the description of the diamond crossing, the speed is limited because of the discontinuities in the running surface. If the crossing angle is small this opening (flangeway gap) can have a significant length and thus a large area in which the wheel flange is not sufficiently guided anymore. To overcome this moveable parts are introduced to fill this gap. It is mostly achieved by making the triangular pointed rails in Figure 29 moveable. It provides a virtual rolling surface and continuous guiding. Other reasons are the reduction of noise and vibrations.

Figure 30 Moveable switch blades in the common crossing part of a diamond crossing, Kijfhoek (NL)[21]

2.4.5 Intermediate rails

Between the wing rail and the point rails, ordinary rails can be found with a normal profile. Rails positioned exactly perpendicular to the sleepers can have normal attachments, but if they are diagonally placed on the sleeper (e.g. from a curved switch blade to the frog) a special seat may be used which allows this placement on the sleeper and normal attachment to the sleepers. The intermediate rails are also known as closure rails, because they close the gap between the moveable switch blades and the frog. The curved rails are exposed to higher forces than in plain track, due to the small curve radius in the curved direction and the unequal support of the longer than normal sleepers, to which they are attached.

2.4.6 Sleepers

The sleepers supporting a switch or crossing are not of the same type as used on plain track (straights and curves), although the same materials are used for new switches and crossings: (tropical) hardwood and concrete. The difference is mainly that they are longer and heavier. This is related to their function in the turnout:
- guaranteeing from intertwined track the appropriate gauge;
- supporting the rails and all other components (check rail, diamond crossings, switch blades etc.) transversally, longitudinally and vertically and transmitting these forces to the ballast;
- guaranteeing the electrical isolation of the two rails.

To these functions has to be added the fact that the resulting forces tend to be higher than on plain track. This results in some sleepers being 3 times as long as a normal sleeper and more than 4 times as heavy.

The metal components of a rail are normally not directly attached to a sleeper, but indirectly and/or elastic. For concrete sleepers, a rail pad is used to damp the impact load, which might crack the concrete (hardwood is much more resilient to this). For hardwood metal seats are used to avoid impressions of the rails into the wood (concrete is much more resilient to this).

For new standard turnouts almost always concrete sleepers are used. For crossings this is more and more the case, but most newly installed diamond crossings with single or double slips are still

---

[21] © Arie van Zon, published with permission by e-mail d.d. 20 June 2008

on hard wood sleepers. Beside this, switches and crossings on wooden sleepers are used if they are installed on a limited ballast height, due to e.g. a tunnel underneath or in any other cases that more elastic support is demanded from the track construction. Sometimes the wooden sleepers are equipped with anchors: steel metal plates rigidly attached to the wooden sleeper, which stick vertically from under the sleeper in the ballast. The main reason is to increase the resistance against transversal loads.

The lengths of the S&C sleepers make them difficult to handle and to put in place. A new development is therefore to connect the sleepers when positioning the partially pre-fabricated switch or crossing at the final site. This is visible in the top of the left picture of Figure 24.

2.4.7 Ballast
On the contrary to the sleepers, the ballast used under and around a switch or crossing is of the same type as the ballast on plain track. It is there to spread the loads from the sleepers towards the support (bridge, tunnel floor or earthworks/embankments) on which the railway is placed. Beside the mechanical function, it is mainly the drainage task of the railway track which is attributed to the ballast.

2.5 Switches & crossings on non-ballasted tracks
A development of the last 30 years in Switzerland but much longer abroad, is to construct railway tracks on a concrete slab, replacing the ballast. This is related to the effect that rail and sleepers tend to move within the ballast. This movement is not bad, as long as it remains an elastic one, which heals after the passage of the trains. However in reality this movement turns gradually into a plastic one, making at, a certain moment, a costly geometrical improvement by means of shifting and tamping of the track, necessary.

This has resulted in the development of track construction type using a concrete slab to replace the ballast, which if properly constructed and in principle should not show any plastic deformation due to the loads exerted on it. Different types exist, from the ones in which the metal components of the switch (rails, switch blades, frog) are elastically attached to the concrete slab, up to the one in which a intermediate concrete sleeper-like block is used, which is placed in a trough in an on-going concrete slab.

2.6 Synthesis
This chapter has shown the classification and typology of switches and crossings, with the curvature and frog angle as the most determining for the use of the switch.
Also the important parts of switches and crossings have been identified and described. Based on the available information from the databases (cf. annex 4, 5 and 6) only the following parts are taken into account for the replacement analyses:
- switch blades and accompanying stock rail,
- the frog,
- the common crossing,
- the check rail.

For the intermediate/closure rail, no replacement data is available. This is also the case for the sleepers, which are almost not subjected to single replacement with SBB – this on the contrary to other infrastructure managers abroad, which do have S&C sleeper replacement as a maintenance procedure. For the ballast, no single replacement (i.e. without replacement of the switch or crossing) is registered.

It is interesting to take into account that this chapter could have been written 20 years ago and the only minor new developments since then would have to be added, e.g. the use of (separable) concrete sleepers and the FAKOP geometry. This also explains the small amount of (academic) references in this chapter. This does not have to be a bad thing; it might also be a sign that the current switch and crossing design is just a very good one. However, when taking into account the amount of degradation and wear and the resulting maintenance and renewal needed –as presented in the next chapters– it remains interesting to conduct research into reducing S&C degradation and wear and thus a smaller maintenance and renewal demand.

3. Degradation and wear of railway tracks

This chapter provides the state-of-the-art on the tear, wear, plastic deformation and degradation of railway tracks. It will describe the phenomena and their causes with a special emphasis on switches and crossings. Also presented is the state-of-the-art of modelling the degradation and wear.

3.1 Introduction

When something is produced, may it be a pump, a refrigerator, a car, an airplane, a train, a stretch of road or railway infrastructure, it is most of the times in its most optimal or very close to optimal condition immediately after production, construction or installation. If it is very close to optimal condition at the moment of installation, it is mostly after teething problems or initial installation settlements that this equipment reaches its most optimal conditions. In this report this situation is referred to as the initial condition Q_i achieved during or just after installation. This quality depends on the quality of the Q_o original device and –if applicable– the quality of the initial placement or instalment. Determining the installation quality is important if an optimal installation quality cannot be automatically guaranteed, e.g. because of conditions not being optimal or a limited amount of time for installation. Although on newly constructed tracks the highest initial quality can be reached due to the fact that the optimal conditions can be created (e.g. optimal and even support, enough time available and working with daylight, parallel tracks closed for traffic), conditions during component replacement on existing S&C or renewals of complete S&C cannot always be optimal (uneven support and no time to improve it, difference between reality and design which has to be resolved in place, working during the night when only one track is available to work on while on the parallel track the train traffic has to continue etc.).

After the installation, train traffic is started with this new equipment. Soon it will show the first traces of quality decrease which are mainly related to this traffic.[22] So apparently, it is the passing trains, which causes the decrease in quality of the plain track and S&C. In §3.2 is explained where these forces come from.

The effect of these forces is a loss of quality. This can happen in various time-dependant ways: linear, progressive, regressive or instantaneous (Figure 32). An instantaneous loss of quality should be avoided, because it could mean a potential safety hazard: it might be impossible to foresee it. How the loss of quality is visible in plain track is explained in §3.2.5.

Knowing how this loss of quality occurs can attribute to the following:
- an improved equipment design;
- an improved installation procedures;
- an improved resistance against this loss of quality;
- an improved forecasting of maintenance and renewal needs;
- a prolonged service life.

Due to the importance of this, §3.4 describes the state-of-the-art on degradation and wear determination.

---

[22] There are some weather related quality decreases imaginable. On tracks which are not in use, within years the first plants and trees will appear. On tracks which are only little used by trains, there won't be that much plants or trees. But due to the long time components will be in place, weather influences might be more important than mechanical quality decreases due to the train traffic.

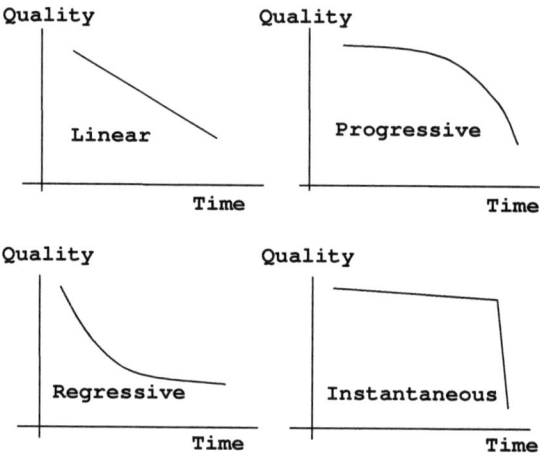

Figure 31 Possible ways of quality decrease

3.2 Theoretical basis: Forces and loads on railway tracks and S&C in particular

As also confirmed by Steenbergen (2008b), it are the forces and loads on railway track causing degradation. This paragraph tries to explain where they come from.

3.2.1 Static loads

Especially the components of the structure involved in supporting vertical train load -the "weight" of the train- are subjected to high stresses. This is mainly related to the railways main advantage: the low rolling resistance caused by the steel-on-steel contact and thus the capability of transporting:
- large amounts of freight by freight trains, with only little traction effort compared to trucks, planes or vessels;
- high amounts of passengers with only a little use of space (e.g. metro or tram systems);
- passengers and light freight at regular or high speeds (respectively conventional railway or high speed lines).

The high loads and stresses in the railway systems can be found for example between a train wheel and a rail, where a maximum static wheel load of nowadays 12,5 tons[23] is transmitted by an area not bigger than that of a coin. The table below shows how this relates to other modes of transport.

---

[23] Value for Switzerland, higher amounts can be found on dedicated freight railways, e.g. in the USA the static load reaches 17.5 metric tons per wheel.

Table 4 Comparison of static loads between transport modes and their respective infrastructure

	Single contact area wheel/infrastructure [cm²]	Load per contact area [metric tons]	Mean stress in the contact area [N/cm²] [24]
Locomotive / freight wagon[25]	1	12.50	122'625
High speed train[26]	1	8.50	83'385
Passenger wagon	1	6.50	63'765
Boeing 777-300ER[27]	[28] 1963	29.00	145
Road truck	[29] 471	[30] 6.00	125
Car[31]	[32] 145	0.50	34
Bicycle	[33] 7	[34] 0.05	70
Pedestrian	[35] 200	0.04	2

This vertical static load is concentrated via the train construction to the wheel/rail surface and spread by the track construction from the wheel/rail surface to the terrain. The high mean stresses as indicated in the table above are an indication of the importance of understanding these forces and stresses when managing (i.e. -in this case- designing, constructing, inspecting and maintaining) the permanent way (i.e. the infrastructure and superstructure), compared to other modes of transport.

The horizontal static loads are only present in case of a train which stops at a curve in which cant[36] is applied or in case of a train standing still at the moment it experiences heavy side wind. They are negligible compared to the horizontal dynamic loads (cf. §3.2.3).

3.2.2 Vertical dynamic loads

As a train starts moving, the forces exceeded by this train on the track tend to increase. The increase of the vertical dynamic load is related to the static load and this relationship can be described with a dynamic impact factor also known as a dynamic amplification factor. Zicha (1989) provided an overview of this as established in the figure below, which shows, that this phenomenon was already known very early in the railway industry.

An explanation of the source for these curves can be found in Annex 2.

---

[24] 1 metric ton = 1000kg = 9810N
[25] Maximum for Europe: 25 tons axle load = 12.5 tons wheel load
[26] According to TSI 17 tons maximum axle load
[27] One of the heaviest charged landing gears of normal airliners: a maximum take-off weight of 350 tons, of which almost all is supported by the 12 wheels of the 2 main landing gears. Source: www.boeing.com
[28] estimated 500mm diameter round contact area $A = \pi \times radius \times radius$. Source: www.boeing.com
[29] based on super-single tyre, 300mm wide and estimated oval contact area length of 200mm. $A = \pi \times (\frac{1}{2} \times length) \times (\frac{1}{2} \times width)$. Source: Vos, E (2003)
[30] source: Vos, E. (2003)
[31] Upper class car: 2 tons total weight
[32] based on a normal 185mm wide tyre and a 100mm long oval contact area. $A = \pi \times (\frac{1}{2} \times length) \times (\frac{1}{2} \times width)$
[33] estimation of the author
[34] 80kg for an average person + approx. 20 kg for the bike
[35] European shoe size 43 (estimation)
[36] or *superelevation* in American English, this is the effect that the outer rail is placed higher than the inner rail in a curve to compensate the lateral/centrifugal forces

3 - Degradation and wear of railway tracks

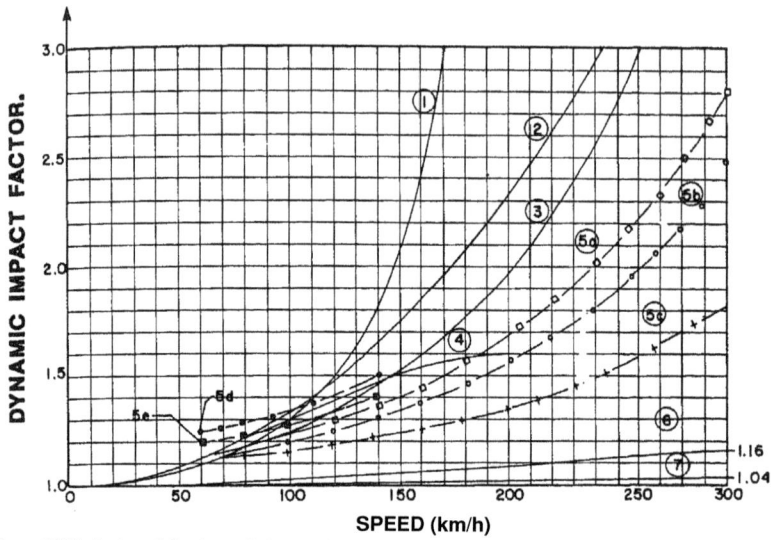

Figure 32 Evolution of the dynamic impact factors for vertical wheel loads acting on the rail [Zicha (1989)]

Based on the origins of the curves in Figure 32 and the article of Zicha (1989), it can be derived that, **beside the static load** and the **speed**, other determining factors for the dynamic vertical load might be:
- Train parameters
    - train construction specificities (e.g. damping properties, centre of gravity)
    - train component condition (e.g. wheel wear; i.e. train irregularities)
- Track parameters:
    - track construction specificities: the designed geometry of the track (e.g. vertical curve)
    - the actual state of the track (e.g. component wear or the condition of the geometry; i.e. track irregularities)

The importance of **train construction specificities** can be in the unsuspended and suspended mass. This has been shown mathematically by Prud'homme (1970), in both theoretical and evaluative terms. The total vertical loads exerted by a wheel on the rail on a straight track can be expressed by the following equation:

$$Q_T = Q_E + 2\sqrt{\sigma^2(\Delta Q_S) + \sigma^2(\Delta Q_{NS})} \qquad (1)$$

where
Q_T total vertical wheel load on the rail
Q_E static wheel load on the rail (kN)
$\sigma(\Delta Q_S)$ standard deviation of dynamic overloads due to sprung mass
$\sigma(\Delta Q_{NS})$ standard deviation of dynamic overloads due to unsprung-mass (kN)

$\sigma(\Delta Q_S)$ and $\sigma(\Delta Q_{NS})$ are evaluated by using the following equations:

$$\Delta Q_S \cong (0.11 \text{ to } 0.16) Q_E \qquad (2)$$

$$\Delta Q_{NS} \cong a \cdot b \cdot \frac{V}{1000} \sqrt{mK} \qquad (3)$$

Lopez Pita (2005) expands this for the Spanish high-speed case with the following values:
- a factor dependent on wheel defects ($\approx$ 0.42)
- b factor dependent on vertical rail defects ($\approx$ 1 - 2)
- V vehicle running speed (km/h)
- M unsprung mass by the wheel (kN)
  - locomotive ($\approx$ 15 kN)
  - carriages ($\approx$ 6 kN)
  - AVE ($\approx$ 10.7 kN)
- K vertical track stiffness (kN/mm)

It is beside the topic of this thesis, but interesting to repeat is one of the conclusions of Lopez Pita (2005), that the formula above indicates, that a reduction of 1 kN of the non-suspended mass (mainly the axle with the two wheels, the discs of the discs-brakes plus the gear box and attached components for powered axles) has a 10 times bigger effect that 1 kN of reduction of the suspended mass (car body and the load in or on it). This reflection is seldom seen in the train industry yet, although at higher speeds the actual dynamic loads do cause serious degradation of the track geometry and wear of the track components, which is explained by Lopez Pita (2005), later in this introduction and in chapter 3 of this report.

Ageing of the trains causes more irregularities of dampers, train wheelband or wheelrim smoothness and thus higher dynamic loads.

The importance of **track construction specificities** can be (as stated) the existence of a vertical curve, where the train (and its passengers/cargo) will be subjected to vertical accelerations when going from a straight track to an inclined track. On a smaller scale, going from large to small, geometrical irregularities can be e.g.
- the changes in subsoil elasticity, e.g. when passing from a track on a bridge to one on soil or vice versa;
- the introduction of cant in a curve during the transition curve;
- rolling mill faults in the rail, which generally occur with a repetition of approximately 3 meters, equal to the circumference of the machines in the rolling mill;
- the rail which is discretely supported by the sleepers (in general every 60 cm), which means a repetitive support-bridge-support effect under the continuous beam formed by the rail;
- and even welding smoothness as confirmed by Steenbergen & Esveld (2006) and Steenbergen (2008a)

As confirmed by the last two articles, even the smallest irregularity of the train track can cause vertical accelerations ($\sim$ dynamic loads) of the train, and thus a bigger effect on the degradation of the track geometry and the component wear.

Ageing infrastructure also tends to have more irregularities, e.g.:
- variable settlement and/or unequal support of the sleepers/ties[37];
- loss of elasticity of components in the track (mainly ballast, rail pads);
- rail surface defects.

Ageing infrastructure therefore also causes more vertical dynamic loads.

---

[37] "ties" is the American English expression.

3.2.3 Horizontal dynamic load

The horizontal dynamic loads are mainly induced by the guiding function of the railway: the relatively simple guiding to stay on the tracks in straights and the forced guiding of the train wheels in a curve. The first one is negligible compared to the latter. The second one is the result of the effect that the moment of inertia of the train "wants" the train to go straight and the curve in the track forces the train in a certain direction away from this. The resulting horizontal component is a dynamic load -it does barely exist when a train stops- exposed on the track by the train wheels. This horizontal dynamic load is (partially) compensated by applying a cant in the track, by putting the outer rail higher than the inner rail and thus tilting the train a little. Because there exists a maximum cant –otherwise the train would tip over if halted in a curve to the inner side– often there is a resulting non-compensated lateral force.

The horizontal dynamic loads are less well understood, compared to the vertical dynamic loads.

Tests have shown that a derailment can occur if the Y/Q ratio over a distance of more than 2 m is greater than 1.2 as confirmed in ORE C116 (1981). For this reason the following value is usually retained as the criterion for safety against derailment:

$$\frac{Y}{Q} < 1.2 \qquad (4)$$

with
Y   quasi-static horizontal wheel load on the rail (kN)
Q   static wheel load on the rail (kN)

Pseudo static approaches are used to determine the maximum forces exceeded on a track, which can be summarized as the not-by-cant compensated lateral force plus the wind force multiplied by a certain dynamic amplification factor as described by Esveld (2001, p.60).
The minimum lateral force which a track should be able to resist is confined in the Prud'homme limit defined by SNCF in the fifties:

$$H_{tr} > 10 + \frac{P}{3} \qquad (5)$$

with
H_{tr} minimum lateral force (in kN) which the track should be able to resist without lateral deformation
P   axle load (kN)

Esveld (2001, p.61) also remarks *"In general the empirical coefficients appearing formula (here '10' and '3') are dependant of the type of track and its maintenance condition"*

We can thus conclude, similar as for the vertical dynamic load, that there is a strong relationship between the horizontal dynamic loads and the maintenance condition.

3.2.4 Static and dynamic loads on switches and crossings

The text above mainly deals with plain track, i.e. straights and curves. But what does this mean for switches and crossings? As explained in chapter 2, a standard turnout has some peculiarities compared to plain track:
- it is an intertwined combination of a straight and a curved direction, so confronted with train loads in several different directions;

- the curve in a switch or crossings in general, depending on the type of S&C, has a small radius compared to normal curves on plain track, thus allowing only low speeds, but with high lateral guiding forces;
- this curve in a switch or crossing has seldom cant applied; both inner and outer rail of the curved direction are at the same height, increasing the lateral force compared to a curve with cant built in;
- beside the small radius, the transition from straight is also extremely short, leading to a high jerk (fast change in acceleration);
- a switch has certain built-in irregularities:
    o sleepers which are much longer than in plain track and of a heavier type, changing the elasticity;
    o specially devised and formed rails or other metal parts which are irregularly supported;
    o rail inclination sometimes reduced to 0 in stead of 1:20 or 1:40 on plain track;
    o a sudden change from rolling contact point from the switch blade to the stock rail or v.v.;
    o a sudden change from rolling contact point from the wing rails to the nose of the crossing;
    o several joints, sometimes electrically isolated.

As stated in the paragraphs above on dynamic loads, it are especially irregularities of trains but also of tracks which are the cause of the dynamic loads. And it are exactly all these irregularities, which can be found back in a switch or crossing.

We can thus conclude that a switch or crossing is exposed to high dynamic loads in both vertical and horizontal direction due to its geometry and/or the built-in high number of irregularities, compared to plain track.

3.2.5 Loads vs. degradation of geometry & track component wear

As the previous paragraph explained, the train load on the track has a certain effect on this track. It will cause degradation of the geometry and tear, wear and plastic deformation of the track components.

The amount of degradation and wear is (again) a function of train _and_ track properties. E.g. a track which is better able to support high axle load, will show less wear than a track which is less well prepared for high axle loads.

Beside this, there is also a relation between track geometry and the state of the track components: components in a bad state will cause more track geometrical degradation, because they are more irregular then components in a good state. This relation is also valid the other way: bad track geometry will in the same negative way affect the degradation of the track materials. The relationships are visible in the figure below.

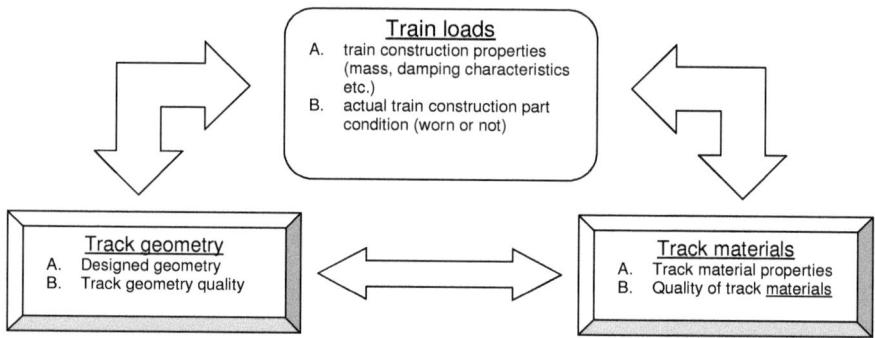

Figure 33 Relationship between train loads, track geometry and the track materials

For various reasons, e.g. to derive maintenance and renewal needs based on the train loads, to determine optimal life-cycle-cost (LCC) policies, and to predict a change in the LCC optimum when changing speed, tonnage or axle load ORE D141 (1982) establishes a method in which the vertical load is considered as decisive:

$$E = k \cdot T^\alpha \cdot P^\beta \cdot V^\gamma \qquad (6)$$

And at the same time the following formula to estimate the change in deterioration

$$\frac{E_1}{E_2} = \left(\frac{T_1}{T_2}\right)^\alpha \cdot \left(\frac{P_1}{P_2}\right)^\beta \qquad (7)$$

where
E   deterioration since renewal or last maintenance operation
k   constant based on rolling contact quality (track geometry, wheel/rail contact area)
T   tonnage (cumulative)
P   total axle load (*static axle load* when V is in the formula, *dynamic axle loads* if not)[38]
V   speed

The factors α, β et γ have been empirically determined in ORE D141 (1982) and other ORE-studies. In ORE D161 (1988) the values in the table below are used:

Table 5 Typical values for the coefficients α, β and γ as a function of the track failures [ORE D161 (1988)]

Phenomenon		α	β	γ
Track material	Rail fatigue	3,0	3,0	1,0 ~ 1,1
	Rail surface defects (e.g. corrugation)	1,0	3,5	1,0 ~ 1,1
	Fatigue of other components	3,0	3,0	1,0 ~ 1,1
Track geometry	Track geometry deterioration	1,0	3,0	1,0 ~ 1,1

Different conclusions from this table are possible, but the most important is that the relative higher importance of the axle load compared to the total load and thus the conclusion that it is better to have a large amount of light axle loads, than a small amount of heavy loaded axles.

---

[38] Thus, in the case of equation 7, it is a dynamic total axle load

What this means for the switches and crossings is that apparently the axle loads are important when determining wear and degradation rates. These are however not exactly known. What is known is the amount of freight trains with a maximum axle load of 22.5 metric tons, relative to the amount of passenger trains which have axle loads of only 11-17 tons for passenger wagons and 22 tons for a locomotive. This can be taken into account when analysing the switch or crossing or its components.

3.3 Modelling the degradation and wear: the approach

A model determining wear and degradation has to be derived with the knowledge that both track and train parameters are involved. To find these influences, in general, there are two different approaches, as pointed out by Selig (1981):
1. an engineering approach and;
2. a statistical approach.

3.3.1 Engineering approach

The engineering approach consists of establishing, by theory and testing, the mechanical properties of all elements that make up the track structure and the railroad vehicles, which run over it. By applying the calculated or simulated static and dynamic loads of the trains and the calculated response of the track elements, permanent deformation can be predicted. This method requires exact description of mechanical behaviour of trains and track.

When it comes to train-track interaction, general models, of which Figure 34 and Figure 35 are examples, are used. They immediately show the challenge: the amount of factors influencing the system, and which thus influence the dynamic loads as described before, is high.

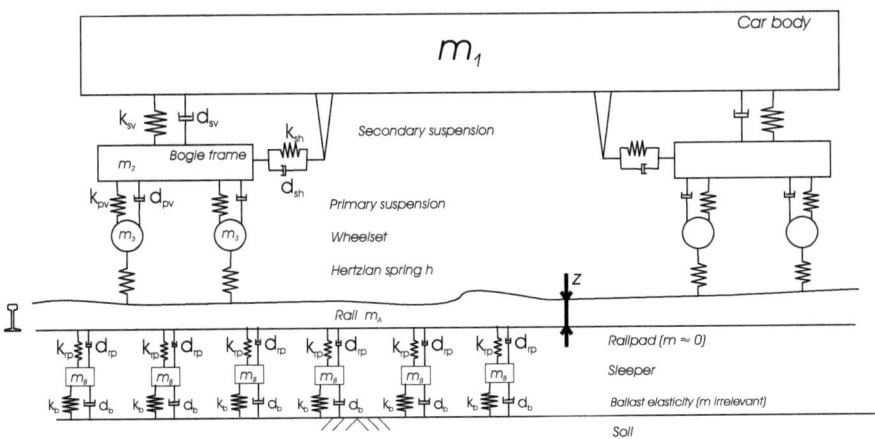

Figure 34 Train-track longitudinal model

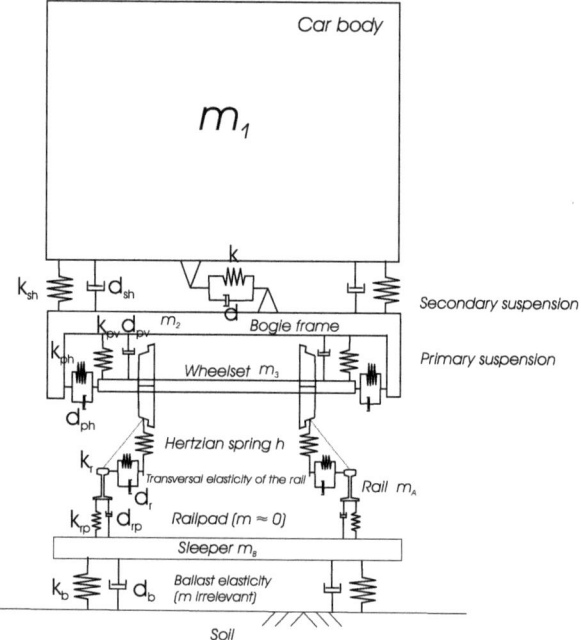

Figure 35 Train-track transversal model

Where
d_b	Ballast damping parameter
d_{ph}	Horizontal primary suspension damping parameter
d_{pv}	Vertical primary suspension damping parameter
d_r	Rail transversal damping parameter
d_{rp}	Railpad damping parameter
d_{sh}	Horizontal secondary suspension damping parameter
d_{sv}	Vertical secondary suspension damping parameter
h	Hertzian spring constant
k_b	Ballast stiffness parameter
k_{ph}	Horizontal primary suspension stiffness parameter
k_{pv}	Vertical primary suspension stiffness parameter
k_r	Rail transversal stiffness parameter (elasticity)
k_{rp}	Railpad stiffness parameter
k_{sh}	Horizontal secondary suspension stiffness parameter
k_{sv}	Vertical secondary suspension stiffness parameter
m_1	Car body mass (fully suspended mass)
m_2	Bogie frame mass (partially suspended mass)
m_3	Wheel set mass (unsuspended mass)
m_a	Rail mass
m_b	Sleeper mass
z	excitation function due to track and train irregularities under load

Taking into account the things mentioned before in this thesis, it is also known that following non-exhaustive list of parameters also have an effect, but are not in the model above:
- Train parameters:
    - the exact length of a railway wagon determines significantly the jerk a tight curve with a short transition curve as can be found in S&C;
    - the distance between the axles (also for the jerk and the transversal accelerations on the rail;
    - the height of the centres of gravity have an influence on the load as exerted on the track.
- Track parameters:
    - rail profile;
    - rail steel type;
    - distance between the sleepers;

3.3.2 Statistical approach

The statistical approach involves –in this case – the analyses of many observations of
- real up-to-date train loads (axle loads, total load, type of trains, direction of trains) over specific switches & crossings (S&C);
- measurements of degradation and wear parameters of S&C or derived effect of this: the maintenance and renewal carried out as a result of this.

For both forms of decrease of switch or crossing (component) quality –geometrical degradation and component tear, wear or plastic deformation– the result of a literature study are presented in the following paragraphs.

3.4 Geometrical degradation

The geometrical degradation of a switch or crossing can show itself as:
- a shift of the complete switch to the left or the right of its designed position due to free movement in the ballast;
- a subsiding leading to unevenness on one side introducing an unwanted cant/superelevation or a subsiding for both rails, leading to a problem in longitudinal direction.

For plain track, the geometrical qualities (and thus its degradation) are expressed in the terms:
- alignment:
    - curvature;
    - longitudinal level;
- cant (transversal level);
- shift;
- twist;
- gauge (distance between the rails).

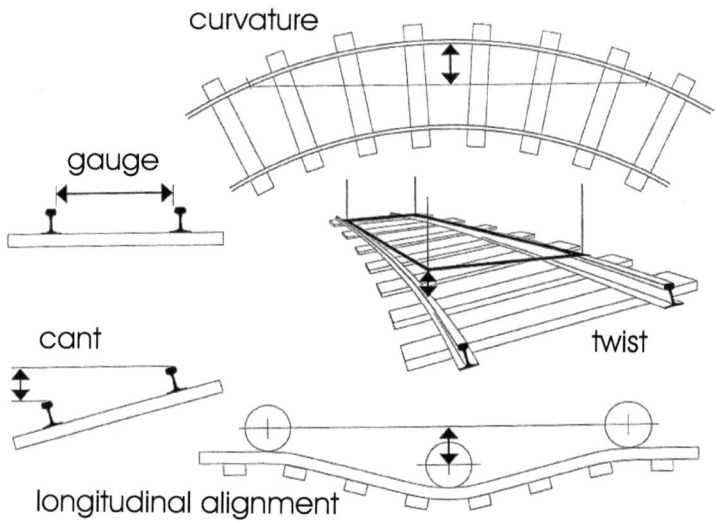

Figure 36 The way in which plain track geometry is expressed

In annex 3 an exhaustive list with all degradation modes is given.

3.4.1 Inspection of geometrical quality

The geometrical quality is for plain track (i.e. straights and curves) measured at high speeds (up to 320 km/h) and expressed in cant, gauge, longitudinal alignment, level, cant, curvature and twist. In Switzerland the whole network is measured twice per year, during spring and fall.
Most measuring trains measure the parameters every 15 cm and store it digitally. There are two ways to express the parameters: absolute and relative. In the absolute way, the measured results are compared to the position the track should be, based on its design. The relative expression is based on the principle that a track might even not be at its designed position, but if the alignment and twist, gauge, curvature and cant do not cause excessive accelerations, the track is also accepted as being well maintained. In principle the relative way was long time used since the absolute measurements could only be done at sight with a clear reference. Nowadays, with the help of GPS and other orientation devices, the absolute measurements can also be carried out from a moving train.

When these measuring trains pass a switch or crossing, the measurements tend to be so full of noise that they are actually thrown away. This is not so strange because of all the irregularities mentioned in §3.2.4 and especially the fact that measuring every 15 cm, there might be quite some changes (switch rail present or not, check rail present or not, nose of the frog).

Therefore some infrastructure managers give for every passage of a switch or crossing by the measuring train a note for the general state of a switch based on the registered accelerations. This however is a very subjective way of determining the exact state of the switch. High accelerations themselves may as well be caused by geometrical degradation as well as by component wear.

It can thus be concluded, that unfortunately no information on actual geometrical situation of the S&C at the SBB network is available.

Still the geometry is nowadays measured with the help of small carts which are pushed through a switch. This mainly regards gauge measurements and groove widths (cf. §2.4.2), but since it is in unloaded state (a measuring cart weighs a couple of kilograms) cannot provide significant results for other geometrical parameters which tend to be only measurable under train load.

3.4.2 State-of-the-art on geometrical degradation modelling

Engineering approach
Studies using an engineering approach –often including laboratory tests– on geometrical degradation of railway track show large importance for ballast properties, since it is only there where settlement can occur: sleepers and rails are in vertical direction much more rigid and the stresses on the soil under the ballast are relatively low to cause big continues after the initial settlements.
Examples of such studies are presented in Mauer (1995a), e.g. Schwab & Mauer (1989), Hettler (1984), Shenton (1975) and the ORE study D117. Currently research is carried out with respect to high-speed track geometry degradation as presented in Schmitt (2006).

Emphasized is the importance of total load and peak loads. The geometrical settlement can be characterised in three consecutive periods:
1. an initial rapid settlement after initial installation, a renewal of the ballast or a tamping action.
2. a relative steady settlement in case of similar train loads
3. an increase settlement and/or increasing unevenness in settlements.

These results can be easily explained by linking them to the state of the ballast. Which in the first period has not been sufficiently packed, but reaches optimal packing under the train loads which results in the settlements of the second period. The fourth period is when the ballast gets polluted through environmental effects or due to fatigue effects in the ballast: the ballast stones being rounded off.

Although these results are only modelled for plain track situations, there is no reason to presume it might be completely different for S&C. However some special effects have to be taken into account.

Statistical approach
As far as the statistical approach for geometrical degradation of plain track is concerned, Meier-Hirmer (2007) provided an overview of studies on this subject, with special emphasis on Veit (2005). Veit shows the progressive degradation of track geometry and how it specially depends on sub-soil and ballast quality (effect of pollution). He also clearly shows the memory-function of the ballast, for which counts that waiting too long with tamping to restore the optimum geometrical quality, might lead to a rapid decrease after the tamping action finally is executed. Also mentioned in this respect should be the studies ORE C116 (1981). In the nineties, the School of Engineering at Queensland University of Technology in Brisbane Australia, conducted research in this topic as presented in Ferreira & Murray (1997), Zhang, Murray & Ferreira (1997, 1998a and b) or Higgins, Ferreira, & Lake (1999). Japanese experience is described in Sato (1995) and Ishida (2001). More recent results from Europe beside Veit (2005) can be found in Larsson (2002 and 2004) for Sweden.

What these studies in general show is a progressive increase in geometrical degradation of the tracks and the fact that this depends on certain parameters. Following parameters are mentioned:
- Train parameters:
    o Speed
    o Axle loads (and exceedance of axle load according to several sources, dealing with heavy haul lines)
- Track parameters:
    o Installed components
    o Sub-soil quality
    o Track maintenance quality

The effect of these is schematized in Figure 37. This shows the progressive degradation and the effect of different speeds with the same kind of trains on comparable track. For the speed is valid V2>V1.

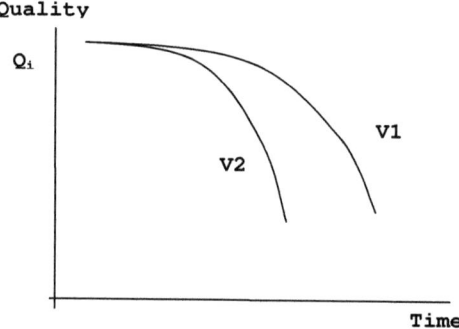

Figure 37 Example of a degradation curve

Due to the higher speeds the dynamic loads will be higher as clearly described by Ferreira and Murray (1997). The same effect as for higher speeds can be seen for higher axle loads, hardly maintained trains (worn wheels, bad suspension), lower sub-soil quality or badly maintained track. Also the installation of unsuited components has its effect. Especially to light components charged with heavy loads tend to show fast degradation.

For specifically switches and crossings, Jovanović & Zwanenburg (2002), Warmerdam (2005) and Scheffers (2007) provide some insight in factors influencing the degradation of the geometry of those S&C. These factors –based on detailed analyses of 1 switch and interviews– they are:
   A. Train parameters:
      a. Axle load
      b. Number of axles
      c. Total tonnage
      d. Speed
      e. Centre of gravity
      f. Maintenance level of specific train components: wheels, primary and secondary suspension
   B. Track parameters
      a. Quality of the frog (the high dynamic (impact) forces of the wheels passing the frog, are suspected to have an effect on other parts in the switch or crossing)
      b. Quality of rail fasteners

c. Sleeper type (wooden types at the end of their life are expected to have a more than average influence on rest life expectations of the complete switch)
   d. Ballast quality
   e. Soil quality
   f. Quality of initial placement
   g. Actual geometry (old switches have sometimes been constructed in place and are not completely conform drawing: this is a cause for uneven ride of the train through the switch, therefore increased dynamic forces and thus a higher maintenance and renewal need during its life)
   h. Maintenance and component renewal policy
C. Usage parameters
   a. Speed
   b. Number of throws
   c. Trains mainly in straight direction or also significant in curved direction
   d. Trains mainly in facing or in trailing direction
   e. If braking or accelerating takes place in the switch

These factors will be taken into account as much as possible in chapter 5 on the explanation of the degradation and wear modelling.
Beside this, the former German state railway company DB has conducted test on the behaviour of their switches on high-speed line presented in Mauer (1995b). They are also presented in Lichtberger (2005, p.379-380) and pictured here in Figure 38.

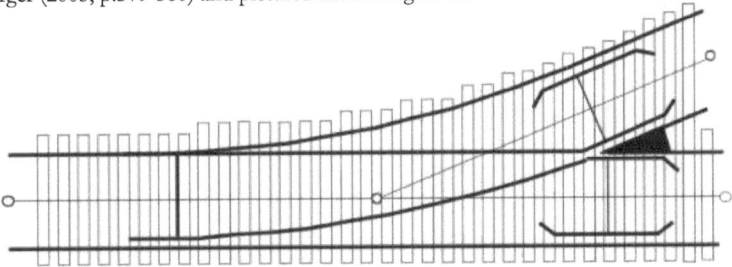

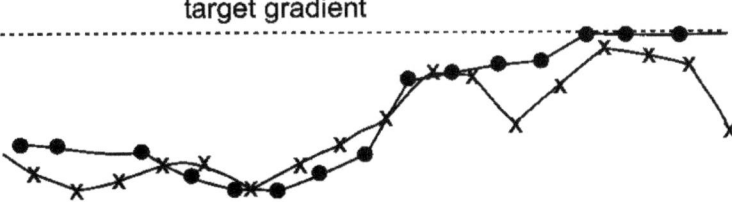

Figure 38 Settlement behaviour of 2 left-hand switches as shown above from Lichtberger (2005)

Lichtberger writes about it:
   The typical settlement is shown in [Figure 38]. A high point develops in the area of the long sleepers, whereas the blade service is subject to the greatest settlement. As switches are passed mainly in the main [straight] track, defects in cross-level occur in the area of long sleepers. These geometric defects are mainly caused by discontinuous supporting surfaces of the switch sleepers and sleeper cavities

We can conclude from this that a switch settles, just like plain track does, but not in the same way: highest settlements can be found in the switch blade area, and lowest at the area where the longer sleepers can be found (just before and around the frog). This is mainly due to the rigidity in the frog area, as described in Kassa & Johansson (2006)

Remains the question how a switch shows geometrical degradation in lateral direction, even more so since the acceleration and jerk are high due to the small-radius-curve and the short transition curve. High use of a switch in the curved direction might therefore lead to more movement in lateral direction of a switch or crossing. In reality, to transfer the forces to the ballast and to increase the resistance against deformation, vertical metal plates are attached under the sleepers. These anchors increase the area in which lateral forces can be transferred to the ballast. Further it must be stated that due to the bigger sleeper type used for switches and crossings, the resistance against deformation in transversal direction is also increased. And die to the heavy construction of all parts, a higher rigidity is introduced in the track construction as compared to plain track. It can thus be proposed that it is only the vertical settlement which leads to geometrical repair works (cf. chapter 4).

3.5 Tear, wear and plastic deformation of components

On the contrary to the geometrical degradation as described in the previous paragraph, the tear, wear, plastic deformation or fatigue can show itself in a multitude of ways on the different components of a switch or crossing.

The metal parts of a switch (switch blade, running rails, frog, check rail) can show
- wear in which material disappears due to the direct (rolling) contact with train wheels;
- tear due to rolling contact or impact loads;
- plastic deformation due to rolling contact or impact loads.

Figure 39 Examples of rail wear. Left: outer rail in a curve. Right: plastic deformation due to overload

Figure 40 Example of rail wear: short-pitch corrugation

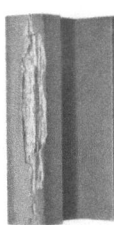 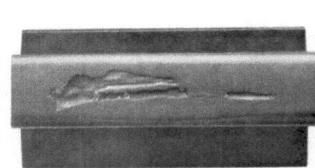 

Figure 41 Examples of rail wear. From left to right: Gauge corner defect, running surface defect and head-checks (last one: junction of St. Benoit (F))

For non-metal parts other things can occur, e.g.:
- wooden sleepers can start to rot away due to weather influences
- concrete sleepers can crack due to the continuous impact loads

A complete list of all failures is provided in Annex 3. This annex is even more exhaustive, because it also includes non-technical failures.

3.5.1 Inspection of tear, wear and plastic deformation of components

When geometrical failures are more or less only properly measurable under load, for finding wear, tear and plastic deformation, more and more sophisticated automatic measurement methods are developed to be used by the measuring trains. This was necessary, since it was not very long ago, that main inspection of wear was on foot. Inspections on foot, are more and more replaced by inspection with the measuring car or from the back cabin of a train. With the help of camera, lasers and ultrasonic devices, rail surface wear, rail profile wear and railhead internal problems respectively are measured.

However –again– not for switches and crossings. S&C are still subject of regular visits of a track inspector at the site to inspect the large amount of different things which have to be measured and inspected (e.g. different gauges, flangeway width, the occurrence of cracks). The visible wear is checked by the track inspector by eye. Wear and plastic deformation is measured with to check is it does not exceed limits. Hardly visible types of wear or wear under the running surface are measured with a handheld ultrasonic measuring device.

Normally, the switches are measured, but as long as the wear, tear and plastic deformation is within limits, it is not noted. This introduces a disadvantage for this study: there exists no single information on the actual degradation or wear during the lifetime of a switch, because it is only written down if the measured values are within limits or not. In the latter case, it is only registered that a certain maintenance or renewal action is necessary, or it is repaired in place (and registered).

Because these measurements are in the track and often traffic cannot be closed down or diverted plus the fact that inspections have to be done during daylight, it remains a dangerous job.
It is only three years ago, that a switch inspection method is developed using sophisticated camera's mounted on a train. At first it was only used to replace the plain track inspections on foot, but more and more they are used for switches and crossings. Video processing allows to automatically analyse the video pictures for irregularities in the track: e.g. rail corrugation, rail joint quality, missing fasteners, sleeper quality or ballast fouling.

3.5.2 State-of-the-art on modelling of the tear, wear and plastic deformation

There exist a certain amount of studies on tear wear and plastic deformation of metal (rail) components under high loads, using both the engineering and the statistical approach.

When analysing the forces and thus causes for wear, tear and plastic deformation from big to small of a train passing through a switch or crossing, we see:
- the short transition curve and small radius and no cant forcing a wheelset (or multiple wheelsets in a bogie) having an effect on the forces exceeded by this wheelset on the track, but this only counts for trains in any curved direction;
- the impact loads at the frog for all directions.

Further in this text, the emphasis is therefore put on the impact loads on the frog and for the rails especially on the effect on the switch rails and stock rail. The problems are mainly related to two metal bodies in non-continuous contact with each other and thus involves tribology.

Engineering approach
Meng & Ludema (1995) discovered, after carrying out a metastudy that 200 'wear equations' exist for wear due to friction of (mainly) metal-on-metal contact. It is serious to then read that Williams (1999) then concludes about that study:
> *However, despite the best efforts of their authors, there is still no way of predicting, with confidence or certainty, the tribological performance of a loaded pair of surfaces, whether dry or lubricated, even if all of their physical and chemical properties have been independently established.*

Especially because this is written with a focus on tribology and wear modelling, this might be regarded as disappointing for this study. However, these were both articles not specifically for a rail. Clayton (1996), Ulrich & Luke (2001), Dang Van & Maitournam (2002), Zakharov & Zharov (2002) and Olofsson & Telliskivi (2003) did put special emphasis on rail wear. It is mentioned that the wear depends on:
- train parameters as mentioned §3.4.2
- the track geometry and the resulting:
    - exact movement of the wheel on the rail: one point contact (wheelband) or 2 point contact (wheelband and flange);
    - angle of attack of the wheel against the rail in the transition curve;
- the fact if the rail is greased or not.
- The rigidity of the rail support (combination of the rail seats, railpads sleepers and ballast, all providing some elasticity)

Also here counts that the degradation is rather progressive: an already degraded rail will provoke higher dynamic train loads, increasing the rail wear.
Kassa & Johansson (2006) present a simulation for train-turnout interactions and plastic deformation of the rail profiles within the turnout. Part of the concluding remarks from the article are worth repeating:
> *The contact pressure in the wheel–rail contact is determined by the magnitude of the contact force and the size of the contact area. It was found that the large contact pressure on the switch rail was mainly due to poor contact conditions. The simulations performed in this study showed that, for a given combination of wheel profile and axle load, the influence of train speed on contact pressure is small. It was also observed that the contact pressure is increased by an increase in axle load. For the nominal wheel profile, the contact pressure generally reached maximum at locations 4–6m from the front of the turnout. However, for the worn profiles, several local maxima appeared at several locations along the switch rail. A combination of high axle load and low train speed resulted in an increased wear index for the worn wheel profile.*

With regard of the frog, it is Fischer et al. (2005 and 2008) who have occupied themselves with a modelling the impact on this crossing. Their analysis and model shows similarities with Steenbergen (2008), who encounters around joints the same impact forces as at the frog.

Interesting to note for this report is the small effect of the speed compared to the state of the wheels and the axle loads.

Statistical approach
As for a statistical approach, there we can find some data on the life expectancy of a rail in North America, related to curve radius.

Table 6 Estimated life of rails in North American Track from Sawley (2001)

Curve radius	Estimated rail life, MGT[39]
Straight	1460
1'750	1050
875	640
580	540
440	510
350	440
290	390
250	380
220	370
190	350
175	330

With these numbers it is clear that in the switches, where radii of 185 meter are common, a high amount of wear will occur. However, at the same time Marschnig &Veit (2006) notes that the wear should not depend on the curve radius since the speed is sufficiently reduced. As can be seen in Table 1, the jerk and accelerations do depend on the switch radius in such a way that they are really different for the admitted speed in a switch. Zhang, Murray & Ferreira (1997) describe beside the geometrical degradation also some wear phenomena, mainly related to the material used for the track construction (rail type etc.). This is however closely related to the high axle loads with which Queensland Rail (the origin of the data for Murray & Ferreira (1997)) on some lines transport minerals in heavy haul trains with much higher axle loads. It is there that the better wear-resistant metals start to count.

3.6 Reflection on degradation and tear, wear, plastic deformation and inspection of S&C

It is interesting to see that for years, the state of the tracks have been expressed in parameters as the geometrical quality (shift, twist, level etc.) and the state of the components (e.g. mm of wear on the rail head), while in reality the train-track interaction force resulting from the geometrical quality and the state of track and train materials are much more important for the comfort and wear. A track may be worn, but if trains only run very slowly at that track still dynamic forces exceeded on it may still be limited. This while at the same time, the most little wear on high-speed lines can have significant effects on the degradation or wear rate. This also means that speed and train type have important influence on the result from the measurement or the calculation of the forces. This idea is not new, already in the 1980's certain measurement trains could calculate the forces on the track from different train types based on the geometrical state as described by Esveld (2001, §16.12).

---

[39] Mega Gross Tonnages, multiply by 100.000 to get tons.

The latest scientific results in this field from plain track (curves and straights) go even one step further: to use a proper mechanical train-track interaction energy management to determine what the optimum maintenance and renewal should be. The first results of these studies, as established in Steenbergen (2008b), show that by employing a proper mechanical energy management as the guideline for maintenance and renewal action, economic gains can be achieved. This might also count for switches and crossings. However, it would be a new PhD study, involving design, maintenance and renewal aspects of switches and crossings.

3.7 Synthesis

This chapter showed that the decrease in quality of a switch or crossing shows itself in 2 ways:
1. Degradation of the geometrical quality;
2. Tear, wear and plastic deformation of the components.

This is caused by the trains passing by, and then, more in particular, the axle load of the trains and the total load on the track. However, since the track loads are also dependant of the track properties, the above mentioned parameters expressing reduction of quality, are also dependant on some track properties. The degradation and tear, wear and plastic deformation of switches and crossings are only since recently measured with measuring trains in some countries, but in Switzerland still by inspections on foot, which results in the effect that no "measurement under load" is available. Especially for geometrical quality this can be misleading: a switch can look perfectly positioned when there is no train on it, but under a passing train it can be completely out-of-specifications. Beside this, the information stored on the measured parameters is only stored if it exceeds limits and when thus a maintenance or renewal action is necessary.

The fact that this information is thus not stored in an appropriate, digital way, means that it is hard to actually model the degradation curve form itself. However, since a maintenance or renewal action is always the result of a certain limit being exceeded, still a model can be derived, based on the information on these maintenance and renewal actions.

The objective of this thesis can also still be reached, because the maintenance and renewal need can still be determined. How that is being done, is expressed in the next two chapters on maintenance and renewal and the methodology respectively.

From the mentioned parameters which have an influence on the degradation or tear / wear / plastic deformation, only the following will be as much as possible taken into account:

Table 7 Parameters taken into account for the analysis

	Parameter	Via
Train	Axle load	% of freight trains tonnage from tonnage database
	Total tonnage	Tonnage
Track	Soil quality	From soil quality data (3 categories)
	Maintenance and component renewal policy	From track category, specifying a maintenance and renewal regime
Use	Trains mainly in facing or in trailing direction	From place in track (S&C database)
	Speed	From S&C database

Following parameters are not exactly taken into account since there is unfortunately no data available:
- Train parameters:
    - axle load;
    - centre of gravity;
    - maintenance level of the trains specific train components: wheels, primary and secondary suspension;
- Track parameters
    - Quality of the frog (the high dynamic (impact) forces of the wheels passing the frog, are suspected to have an effect on other parts in the switch or crossing)
    - Quality of rail fasteners
    - Ballast quality
    - Quality of initial placement
    - Actual geometry (old switches have sometimes been constructed in place and are not completely conform drawing: this is a cause for uneven ride of the train through the switch, therefore increased dynamic forces and thus a higher maintenance and renewal need during its life)
- Maintenance and component renewal policy
- Usage parameters
    - Number of throws
    - Trains mainly in straight direction or also significant in curved direction
    - If braking or accelerating takes place in the switch

Results will be separately presented for specific parameters:
- Switch or crossing type
- S&C angle / curve radius
- Sleeper type

4. Maintenance and renewal of switches and crossings

The possible maintenance and renewal actions on switches and crossings, the maintenance and renewal policy and the maintenance and renewal optimisation form the contents of this chapter.

4.1 Introduction

The degradation of the geometry or the tear, wear and plastic deformation cannon continue unlimited. At a certain moment the degradation or wear of track reaches a limit on which comfortable and/or even safe train traffic cannot be guaranteed. A maintenance or renewal action is then necessary. The choice for which maintenance or renewal action has to be carried out and the appropriate time to do this is the subject of a long-time maintenance and renewal policy optimisation process. But in principle the two things are complementary. They are even related: It is quite useless to replace a worn rail with a new rail on a track which is in a bad geometrical state. Similar inefficient is improving the track geometry, while at the same time a worn rail might provoke such dynamic forces on the track construction that the proper geometry will not last long. Also the component choice should be well chosen, relative to the track geometry demanded (no heavy rail on a tight curve as normal on a tram line). These relationships are expressed in the Figure 42 below which has strong links with Figure 33.

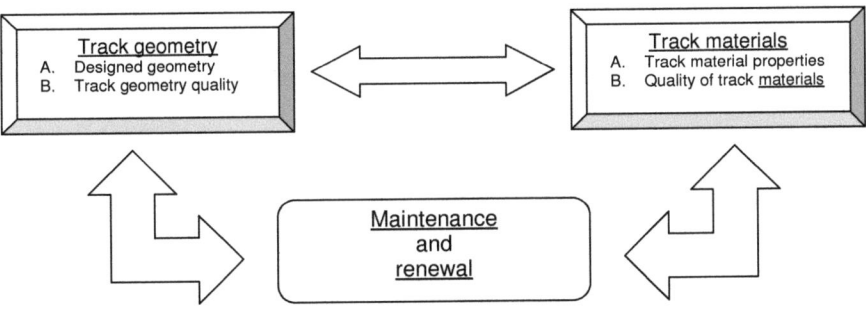

Figure 42 Relationship between track geometry, track materials and the maintenance and renewal works

A maintenance action is here defined as a repair action where there is not actually a component replaced. A renewal action is where a switch or crossings component or a complete switch or crossing is replaced.

For every degradation type, tear or wear there exist a certain maintenance or renewal action or sometimes there is exist even a choice between them. E.g. wear on the frog can be repaired by welding a new piece of metal on (a difficult repair work, which can only be carried out under special circumstances), or by replacing it completely (a heavy job, but the risk for doing something wrong is significantly reduced compared to welding).

Sometimes, a maintenance action might be the normal reaction to certain degradation or wear types. But if the frequency of these maintenance actions is increasing, it might be that a replacement is a better solution. E.g. when the geometry degrades rapidly, also every time after a tamping action, it might be that the ballast quality is simply too bad and that either the ballast or the switch completely needs to be replaced (ballast replacement under a switch is not an easy task and much easier to do if the switch or crossings is not in place).

The fact that a maintenance or renewal action on a switch or crossing (component) is necessary is determined by the inspections on foot as described in the previous chapter. Most of the time already a couple of months before a components reaches its wear limit, it is known that a maintenance or renewal action is necessary and it can be planned ahead. There are however some constraints. Due to a limited budget, the infrastructure manager responsible for planning the maintenance or renewal has not unlimited personnel, machines and spare parts. Beside this, there is not that much time to carry out this maintenance on very highly utilised train and metro networks. Therefore he has to find an optimum on carrying out the works in the limited amount of time that on a certain stretch the train traffic is halted or rerouted.

There are several options to optimise this maintenance policy. Some of them are standard optimisations, widely used in several different industries, e.g. condition-based maintenance total-productive maintenance or reliability-centred maintenance, all with their specific drawbacks and advantages. For most of these policies or strategies a thorough understanding of the assets is necessary. Prediction of wear trends or continuous automatic condition determination is sometimes a necessity on which the complete strategy relies.

The maintenance and renewal actions on a switch or crossing are in a non-exhaustive list gathered in annex 4 with this report.

4.2 Maintenance actions on S&C

There are several maintenance actions that can be carried out on a switch as a result of mechanical wear, they are summarized in Table 8.

Table 8 Maintenance actions on a switch or crossings

Component	Maintenance action
S&C geometry	Correction by tamping
Moving parts	Greasing or adjustment
Metal parts	Grinding
Frog	Welding

First of all there is the correction of the geometrical degradation by way of tamping. In this case a tamping machine will position itself on top of the switch or crossing and –based on measurements– will position the switch or crossing at its proper horizontal and vertical position. Because the ballast will due to that not support the switch sleepers anymore, the tamping machine has to tamp the ballast under the sleeper, to supply sufficient support for the switch. The tamping machine is therefore equipped with vibrating "packers" entering the ballast and squeezing the ballast under the sleeper.

Another regular maintenance action is the greasing of the moving parts. Nowadays more and more switches use carbon-fibre sliding plates or rollers which do not have to be greased, it is therefore not a standard maintenance action anymore and sometimes carried out together with the regular inspections of a switch. Because there exists no data on the greasing, this work is ignored in this report.

As far as the metal parts, subjected to train loads, is concerned, they will wear due to this. Sometimes this happens uneven, which can be repaired by grinding off the burrs. Sometimes also other wear effects occur, e.g. rolling contact fatigue (head checks, squats, tache ovals or shelling) or gauge corner cracking, which, if discovered in an early enough stage, can also be stopped by grinding. More developed rolling contact fatigue or gauge corner cracking makes a complete replacement of the rail necessary.

Although all metals can be welded, only on the frog or the common crossing, welding is carried out on a regular basis: for a normal rail it is easier to just replace the rail and for the switch rail with accompanying stock rail, the weld itself tends to be too difficult because of the continuous changing area over the length of the switch rail. A frog or common crossing however is a big piece of metal connected to 4 rails, of which complete replacement is also a major work.

With regard to the database, only:
- tamping;
- welding on the frog;
- grinding

are registered and therefore analysed later in this report.

4.3 Renewal actions on S&C

Beside the maintenance actions, there are also the complete renewals, as presented in Table 9.

Table 9 Renewal actions on a switch or crossings

Renewal action
Switch rail with accompanying stock rail
Frog
Intermediate rails
Check rail
Sleepers
Ballast
Complete switch or crossing renewal (incl. ballast)

The major one of these works is the renewal of a complete switch or crossing. At least one big or two small cranes are needed to lift the old switch or crossing and put in place the new one, which is nowadays often delivered in only three to four complete plug-n-play pieces.
Also for the replacement of other parts a mini-crane is used.

Especially for the metal parts, replacement involves often the cutting up and later welding-together of metal parts. This is an renewal work which requires expertise, since the smallest bad-positioning can lead to a reduction in the expected life, due to the vibrations it will cause described by Steenbergen (2007).

Although sleeper and ballast are mentioned here, they are seldom replaced separately on the SBB network. Some countries however see the replacement of only a couple of worn sleepers as a normal renewal policy. It remains however a difficult job to do without destroying the support function of the ballast.
Ballast replacement under a switch or crossings is something which since recently can be carried out by ballast replacement machines. They dig the ballast from under the switch or crossings by using a long chain which will pass under the switch and which will remove the old ballast. New ballast is later dropped from above between de sleepers. Also this is a difficult task, especially to arrive at a proper homogeneous support. Sometimes the switch is also removed before ballast replacement and placed back afterwards. This description makes clear that also this is not an easy task. This is the reason that it is a job not often carried out and also not carried out with SBB. Ballast is in Switzerland renewed at the moment a switch or crossing itself is renewed. This renewal is the result of the fact that wear of multiple sleepers in combination with one or more other component (switch rail, check rail) reaching its wear limit.

The renewal of ballast without switch or crossings renewal might however get a more important role in the future, since more and more switches and crossings are introduced equipped with concrete sleepers. They tend to have a longer service life, even longer than the ballast lasts.

With regard to the database following replacement are taken into account:
- complete switch or crossings replacement
- switch rail with accompanying stock rail
- frog
- check rail

4.4 The optimal moment to carry out maintenance or renewal

Maintenance and renewal would be most simple if an infrastructure manager could measure the state of its assets in (almost) real time and if the amount of people, equipment, spare parts and possession time is unlimited. All these parameters have however a limited availability, mainly budget restricted. These, combined with other reasons, lead to the fact that maintenance and renewal cannot be carried out at the optimal moment: just before a safety limit Q_s will be reached. This is the reason that maintenance and renewal is in general planned and carried out before the end of life is really reached (time related) and before the minimum quality is reached (quality related), up or around the point Q_{min}, also known as an "intervention limit".

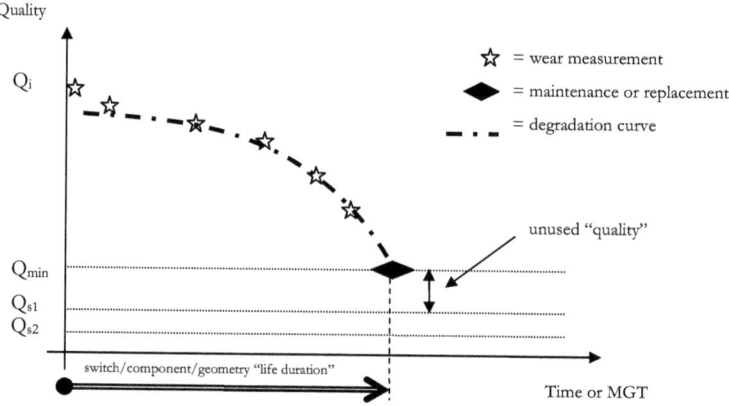

Figure 43 Quality during the life of a S&C or its component

This leads to a certain amount of "quality", which remains unused to carry out its purpose, and is thus a loss of efficiency.

For a high amount of parameters (e.g. for the geometry: shift, twist and level; for the wear: the rail head wear) the safety limit Q_s depends on the train speed. In Figure 43 it the counts for Q_{s2} < Q_{s1} because for the respective speed limits count V2<V1.

One task of the infrastructures manager should be to reduce the gap between the safety limit and the intervention limit as much as possible.

4.5 Maintenance and renewal policy

4.5.1 Description

The maintenance and renewal policy can be described as the repartition of the budget between maintenance and renewal. As described before in this chapter, for a lot of works, a choice can be made between those two, on different level e.g.
- to weld a new part in a frog (=maintenance) vs. to replace it completely (=renewal);
- on a much bigger scale: to replace every time only one component of a switch or crossing: ballast, sleepers, switch rail, frog, check rail (=partial renewal ≈ maintenance) vs. replacing at a certain moment a complete switch to be able to introduce new technology (concrete vs. wooden sleepers) or a change in lay-out of the infrastructure (=complete renewal).

In search for the optimum there are some constraints. The most important of these is that all objects in railways have a very long lifespan and the results of this is that:
- decisions regarding infrastructure, made now, will not have an immediate and impressive effect;
- the replacement rates are low (if all switches and crossings remain in track on average 25 years, only 4% of the switches and crossings will be replaced every year) thus also possibility of introduction of new technology;

Therefore also, changing from a more maintenance to a more renewal focused policy has to take into account some time and often a budget increase as described in Rivier, Putallaz & Zwanenburg (2005).

Related to this are subject are life expectancy and substance as explained below.

4.5.2 Life expectancy and substance

According to Rivier & Hofmann (1994) a maintenance and renewal policy, should have as its main goal to achieve the lowest costs at the long term mainly by learning from all the previous experiences and continuous optimisation.

One of the simplest theories regarding this optimisation problem is that the infrastructure of a heterogeneous and "adult" network is an infrastructure:
- of which the components (tracks, switches and crossings, signals, catenary, etc.) have different age;
  - because they are constructed at different moments in the past;
  - because they have been maintained or replaced at different moments in time;
- subjected to different loads on different part of the network, resulting in different degradation rates in time.

If the two points above are true, for a heterogeneous network counts that the average age of infrastructure components should be approximately half the average life expectancy of these components as described in Rivier & Hofmann (1994). These life expectancies can differ due to the different degradation and wear rates infrastructure components have.[40]

---

[40] This only counts for a heterogeneous network. A homogeneous network or rail line has very different demands. A newly built line will for the large part have peaks in its maintenance and renewal demand. E.g. the tracks of the high-speed line Paris-Lyon needs replacement of the ballast on its whole length (400km) because the end of life has been reached as described in Rivier, Putallaz, Zwanenburg et al (2005). The necessary investment is higher than the original installation cost of the newly constructed track.

If this is true, than also all the following theories are valid:
- the annual renewal rate of components equals

$$\frac{\text{life expectancy}}{100}\%\qquad(8)$$

- life expectancy decreases if existing infrastructure is subjected to higher loads and v.v.;
- life expectancy increases when installing better equipment, if the load stays the same.

When an infrastructure component, e.g. a rail, has a life expectancy of 10 years, in a heterogeneous network, 10% of the total rail length should be replaced annually.
The challenge for the infrastructure manager of a heterogeneous network, described by Rivier, Putallaz, Zwanenburg *et al.* (2005) and Putallaz (2007), is to find this optimum between the above presented life expectancy and annual replacement and renewal also known as *conservation* of the *substance*, taking into account economical, political and organisational constraints.

The infrastructure manager does have a couple of tools to optimise maintenance and renewal. One of those is to apply different policies fro different lines, based on the traffic. The UIC uses for this a division in 7 categories as described in UIC leaflet 714.R. This division calculates a fictive tonnage, taking into account the actual tonnage (train load) and speed, including a division between passenger and freight wagons and locomotives. Some countries still use an older version of this leaflet, making a division into 9 track categories as explained in Rivier, Putallaz & Zwanenburg (2005).
Switzerland uses its own division into main tracks (German abbreviation: *HG* from *Hauptgleise*, French: *VP* from *Voies Principales*) and secondary tracks (German: *NG – Nebengleise*; French: *VS – Voies Secondaires*). Both categories are subdivided into 3 subcategories of which the main characteristics are summarized in the

Table 10 Track category description [SBB]

Category	Principal characteristics
HG1	Main tracks with a daily cumulated tonnage >30.000 tons (previously >25.000 tons), track allows for high speeds
HG2	Main tracks with a daily cumulated tonnage between 10.000 and 30.000 tons (previously 10.000 and 25.000 tons)
HG3	Main tracks with a daily cumulated tonnage < 10.000
NG1	Secondary tracks with daily cumulated tonnage comparable to HG1 and HG2, access tracks to marshalling yards and freight yards
NG2	Main sidings to let trains overtake each other and track leading to platform tracks of terminus station (German: *Sackbahnhof/Kopfbahnhof*; French: *cul-de-sac*)
NG3	All other tracks not within NG1 and 2.

4.6 Maintenance and renewal (policy) optimisation

4.6.1 Introduction

In the railway industry, the assets (e.g. trains, tracks, stations, bridges, tunnels) have long life durations, e.g. for trains minimal 30 years and for tracks often even longer, before they are replaced. An optimal maintenance and renewal strategy is therefore almost automatically a strategy which focuses on the lowest costs on the long term. This has resulted in several Life Cycle Cost (LCC) calculation methods summarized in Meier-Hirmer (2007). The input data of several of these models are expenditure trends in the past, which are then extended as a trend for the future. More sophisticated models take the actual maintenance and renewal actions into account and are able to attribute a cost to it. A disadvantage of both models however is the fact that they do not take into account if these expenditures were enough to guarantee a certain substance, i.e. if the overall quality of the assets changed.

The more sophisticated models also allow introducing more easily uncertainties; changes in the use of the railway tracks (e.g. more trains, faster trains, heavier trains) and relationships between for example soil quality, track loads and thus the expected degradation. This is possible, because a strong relationship between these parameters is proven for plain track, on which the above mentioned LCC models focus. This proof is not (yet) available for switches & crossings. Expert input via interviews on this subject however exists in Jovanović & Zwanenburg (2002) and Scheffers (2007), which is also mentioned in the previous chapter and summarized in Table 11: more trains, higher axle loads, higher speeds or more trains in the curved direction of a switch or crossing cause more wear, similar for bad soil causing high deterioration rates. The "condition" category however also indicates that the state of certain S&C-components also determine the degradation of other components, also confirming the relation between state of geometry and state of materials as described in the introduction of this chapter.

Table 11 Parameters influencing wear and degradation of switches & crossings [Jovanović &Zwanenburg (2002) and Scheffers (2007)]

Category	Parameter	Estimated influence
Loads	Tonnage	High
	Maximum axle load	High
	Main direction of use (straight/curved, towards the point or not)	High
	Number of axles	Above average
	Area of acceleration/deceleration	Average
Constants	Soil quality	High
	Initial condition of placement	High
Condition	Geometry	Above average
	Ballast	Above average
	Sleepers	Above average
	Attachments	Average
	Switch diamond/frog	Average
Maintenance	Quality	High

Apparently, according to experts, the condition of an S&C-component can influence the wear rate of another component in the same switch, and a bad geometry can be the cause of higher wear rates of switch components. This has to be taken into account when optimising maintenance and renewal (policy).

4.6.2 State of the art on maintenance optimisation

With regard to maintenance optimisation of S&C, only the work of Marschnig & Veit (2005) and Veit (2005) have established something that goes in that direction. They present it as part of larger scale study into a LCC-optimised maintenance and renewal strategy for the Austrian Railway. The differences in the approaches to find the optimum are presented in Zoeteman & Veit (2003). Where, for the Dutch case, only a single S&C replacement optimisation is presented. Marschnig & Veit (2005) however present the result of an optimum renewal analysis, of which the main conclusion is that in almost all cases a switch or crossing on concrete sleepers with 60kg/m rails is the optimum solution. Only on tracks with lower loads (leading to much less wear and thus lower rate of return when taking the traffic into account with the profitability calculation), or on tracks with very bad subsoil (leading to a very low life expectation of all track components), lighter rails and wooden sleepers should be used.

Zoeteman (2004) on LCC takes also the reknown RAMS[41]-method into account and describes it in detail. This method describes the four RAMS properties of a component or a system, the latter thing being a switch in this case or even a couple of them. This notion is introduced also in Warmerdam (2005), but it remains very basic with regard to the RAMS of the non-electrical components of a switch or crossing. Failure of a mechanical switch component is actually not a failure, in the sense of a crack, break or whatever, but the non-compliance with maximum wear limits. They introduce insecurity in the result, because there is not only a maximum wear limit, but also an intervention limit. Between those two limits, trains can still run – the track is still available, but an intervention has to be planned. Failure is therefore less black and white as presumed and application of the RAMS as described in the European Norm EN 50126, difficult to apply at the moment that exact data lacks.

On a more abstract level, research into general maintenance and renewal optimisation models has been widespread, although Dekker (1995) mentions that a high amount of publications is combined with only a small amount on actual applications. According Dekker (1995) that is mainly due to the tailored-made solutions found for every situation. The reason for this is summarized by Wang (2002): the large amount of choices that have to be made when searching for an optimal maintenance policy. For example a reliability-centred maintenance optimisation for S&C will find a completely other solution for the situation on the left in Figure 44, where no redundancy exists for turnout A for trains from 1 tot 2, as for the situation on the right in the same figure. In the last situation, turnout B may fail, since via turnout C trains can still continue from 1 in the direction of 2 as long as this failure does not endanger a safe passage of the train. For several reasons this single-component (on S&C level) notion is not continued here:

- Reliability of S&C is mainly related to their main functional failure. Analyses have shown this is mainly related to the failures of the actuator/motor and heating in the winter as described by Jovanović & Zwanenburg (2002). These are not subject of this report.
- The safe passage can seldom be guaranteed after mechanical failure of a component as mentioned in the report, e.g. rail breaks, wear which passed the limits in which a switch or crossings can be used without chance of derailment;
- Data on S&C redundancy is not available.

---

[41] An abbreviation for Reliability, Availability, Maintainability and Safety/Security – a philosophy which orginates from electrotechnical engineering

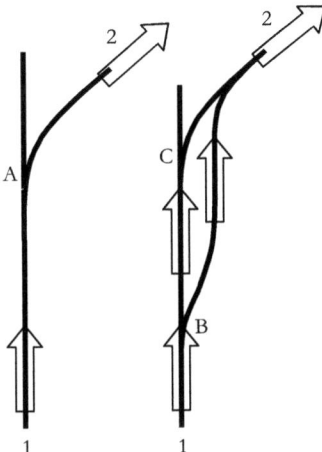

Figure 44 Quality during the life of a S&C or its component

What remains on switch level are the multi-component maintenance optimisation models, as used if a switch or crossing is analysed by itself. Important in this case are the interactions between the components. Thomas (1986) distinguishes economic, structural and stochastic dependence. Nicolai & Dekker (2007) define and describe these three as follows:

> (...) *economic dependence* implies that the cost of joint maintenance of a group of components does not equal the total cost of individual maintenance of these components. The effect of this dependence comes to the fore in the execution of maintenance activities. On the one hand, the joint execution of maintenance activities can save costs in some cases, e.g. due to economies of scale. On the other hand, grouping maintenance may also lead to higher costs, e.g. due to manpower restrictions or [it] may not be allowed.
>
> *Structural dependence* applies if components structurally form a part, so that maintenance of a failed component implies maintenance of working components, or at least dismantling them.
>
> *Stochastic dependence* occurs if the condition of components influences the lifetime distribution of other components. Synonyms of stochastic dependence are failure interaction and probabilistic dependence. This kind of dependence defines a relationship between components upon failure of a component. For example, it may be the case that the failure of one component induces the failure of other components or causes a shock to other components.

Interaction matrices can be made for the economic (E) interactions and the maintenance and renewal actions (Table 12)and another one for the structural (U) or stochastic (O) dependences of the components (Table 13) of a switch or crossing.

Table 12 Economic (E) maintenance and renewal interaction

		maintenance			renewal			
		tamping	welding on the frog	grinding	complete switch or crossings replacement	switch rail with accompanying stock rail	frog	check rail
maintenance	tamping	E						
	welding on the frog	-	E					
	grinding	-	-	E				
renewal	complete switch or crossings replacement	E	-	-	E			
	switch rail with accompanying stock rail	-	-	-	E	E		
	frog	-	-	-	E	E	E	
	check rail	-	-	-	E	E	E	E

The economic interactions are mainly present, if maintenance and renewal costs are different when combining several of these works. This might only be the case if:
- set-up or installation costs have to be payed only once;
- the same machine and expert personnel can be used more efficiently.

The value of possession time[42] has not been taken into account, also because different sources explained the relatively low amount of this on the Swiss network, e.g. Putallaz (2007).

Table 12 shows that all similar works (tamping with tamping etc.; the central diagonal in the table) have an economic dependence. And so does the installation of a new switch or crossings with tamping, since a tamping machine has to be available when installing the new switch or crossing. A switch rail with accompanying stock rail renewal, frog renewal or check rail renewal combined with another also leads to total cost differences since these works are carried out with similar equipment (welding, machining, small crane etc.) and expert personnel. All of the indicated economic dependencies are positive[43]: costs are reduced when the equipment or personnel can be used multiple times on or close to the worksite for different or equivalent works as indicated with an E.

Welding on the frog is quite a specific job and not related to any other job through equipment or personnel it is also different from making an alumino-thermic or electric weld to join two rails together or to attach a new frog or switch rail to existing rails.

---

[42] the time that a track is specifically out-of-order because of maintenance works
[43] the definition of positive and negative are similarly defined by Nicolai and Dekker (2007)

Table 13 Structural (U) and stochastic (O) dependence S&C components

	switch blades and accompanying stock rail	frog	common crossing	check rail
switch blades and accompanying stock rail	-,-			
frog	-,-	-,-		
common crossing	-,-	-,-	-,-	
check rail	-,-	-,O	-,O	-,-
geometrical condition	-,O	-,O	-,O	-,O

As can be seen in the Table 13, not a single structural relationship exist between the components mentioned here: when replacing any of them, no other has to be removed temporarily. This even counts for components not mentioned here: ballast, sleepers, the switch actuator/motor or the switch heating. For the stochastical relationship it is different. As mentioned in the introduction of this chapter, the geometrical state of a switch or crossing can have an effect on the tear, wear or plastic deformation rate of the components, with the opposite also true: tear, wear or plastic deformation causes a more uneven running surface, thus provoking higher loads on the ballast and thus faster geometrical degradation. The result can be seen at the bottom row of Table 13: all component conditions are stochastically related to the switch geometry and the opposite way around. This also counts for the check rail vs. the frog and the common crossing. The check rail has an important function in avoiding the train wheel to take the wrong direction at the frog or common crossing. However a worn check rail has an effect on the frog, since it increases the play of a passing wheelset. The effect is similar if this check rail protects a common crossing.

This stochastic dependence increases the complexity of multi-component maintenance and renewal optimisation through modelling. However for this study some presumptions can be made which overcome these problems. They are presented in the methodology.

On component level, a single component maintenance optimisation is possible, but as Dekker (1995) mentions, therefore exact degradation trends should be known, which is unfortunately not the case, since they are not registered. Also, this is part of the operational works, which lies out of focus for this study.

4.7 Synthesis

It's a pity to see that rather a small amount of information on both degradation, tear / wear / plastic deformation and the actual maintenance and renewal is available. With regard to this, only Veit (2006) comes close to a maintenance plan, but it is unclear where this plan stems from and what was the basis for this maintenance plan. Also its conclusion that with a 80.000 daily tonnage little maintenance to switch rail is necessary is rather an underestimation: experts count it beforehand as one of the most maintenance intensive tasks and regard the switch rail with accompanying stock rail as the part to be replaced earliest.

Therefore, the synthesis for this chapter limits itself to establishing the links between the different maintenance and renewal works and the parameters that might have an influence on the need for this maintenance and renewal works as defined by experts and which are available for analysis.

Combining the information from previous chapters with the information in this chapter, results in following combinations that can be established, on which the methodology as explained in the next chapter is based.

Table 14 Input for the methodology: maintenance

degradation or tear / wear / plastic deformation	Maintenance action investigated	Parameters
Degradation of geometry	Tamping	• Tonnage
Rail wear / frog wear	Grinding	• % of freight trains from tonnage database
		• Soil quality
		• Main train direction
Shelling of the frog	Welding on the frog	• Speed
		• Sleeper type

Results for the maintenance needs will be separately presented for specific parameters:
- Switch or crossing type
- S&C angle / curve radius

Table 15 Input for the methodology: renewals

Renewal action of	Parameters
Complete switch or crossing	• Tonnage
	• % of freight trains from tonnage database
	• Soil quality
	• Main train direction
	• Speed
switch rail with accompanying stock rail	• Tonnage
	• % of freight trains from tonnage database
frog	• Soil quality
	• Main train direction
check rail	• Speed
	• Sleeper type

Results for the renewal actions will be separately presented for following specific parameters:
- Switch or crossing type
- S&C angle / curve radius

5. Methodology & theoretical basis

This chapter repeats the available data as described in the synthesis of previous chapters. It describes then the methodology applied to this data to derive the degradation and wear models, of which the result will be presented in the next chapter.

5.1 Available data & database description

The main, abstracted, input information are the databases with:
1. the description of the properties of all the S&C
2. the description of all maintenance carried out on S&C
3. the description of all component renewal actions carried out on S&C
4. the track loads of the last 50 years
5. some other databases

Database 1, 2 and 3 are especially for this study abstracted from the DfA[44] database of the Swiss Federal Railways (SBB). Database 4 is derived from track usage information, also from SBB.

The databases are described below.

5.1.1 Switches and crossings database, general description

A description of all switches and crossings on the SBB network (situation of march 2005) can be found in the specially for this study composed database WEITABL1:
- a unique number (unique for the whole database, especially introduced by the SBB for this study and now also internally used)
- place
    - geographical indication (a so-called *Betriebspunkt*; most of the time the name of a station or a nearby city)
    - a number (1, 2, 3A, 3B etc. which are used every time again for different geographical indication)
    - track number
- type (standard turnout, symmetrical turnout, diamond crossing etc.)
- rail profile
- curve radius (or radii in case necessary)
- straight type or both directions curved
- crossing angle (1:7, 1:9 etc.)
- sleeper type (wood, steel concrete etc.)
- direction (left-hand switch, right hand switch etc.)
- type of switch rail (spring- or articulated switch rail)
- track category (HG1, HG2, HG3, NG1... etc.)
- different maintenance schedule related information
    - maintenance groups (2, 4 or 6-year periods group)
    - drawing indicator
- track gauge (normal gauge, narrow gauge, or a combination of the two)
- lengths of the different directions

---

[44] DfA = [German] Datenbank für feste Anlagen (Database for immobile assets). It contains all information on the current situation of the track and wayside assets (tunnels, bridges etc). It also includes the maintenance and renewal works carried out on those assets.

- the applied cant (if placed in a curve)
- year of placement
- amount of years a previous switch was in place
- recycle indication (if some parts are reused: sleepers, rods)
- several indicators for the presence of special parts:
    - special sleepers heads,
    - special construction parts and if spares of them exist
- ballast type and quality
- thickness of sub-layer
- information on the welding
    - date
    - rail temperature
    - company responsible for the welding
- switch heating type and operation control (automatic or not)
- switch motor or actuator type and operation control
- switch interlocking type and number
- information on property rights (SBB is not 100% owner of all S&C)
- information on maintenance and renewal responsibility (SBB is not 100% responsible for the maintenance and renewal of all the S&C on its network)
- provisions on
    - complete renewal
    - partial renewals or previewed maintenance works

A full description of the database can be found in annex 4.

5.1.2 Maintenance works on S&C database

In the database WEITABL2 are registered the maintenance works as single works on a switch or crossing for the period 1997-2005. For every maintenance work is registered:
- the unique number for a switch or crossing (unique for the whole database, especially introduced by the SBB for this study and now also internally used)
- place
    - geographical indication (a so-called *Betriebspunkt*; most of the time the name of a station or a nearby city)
    - a number (1, 2, 3A, 3B etc. which are used every time again for different geographical indication)
- the year the work has been carried out
- the type of work:
    - tamping
    - screw hole renovation
    - welding on the frog
    - grinding
    - checkup

A full description of the database including some results on the data mining can be found in annex 5.

5.1.3 Renewals on S&C database

In the database WEITABL3 are registered the renewal works as single works on a switch or crossing for the period 1997-2005. For every renewal work is registered:
- the unique number for a switch or crossing (unique for the whole database, especially introduced by the SBB for this study and now also internally used)
- place
    - geographical indication (a so-called *Betriebspunkt*; most of the time the name of a station or a nearby city)
    - a number (1, 2, 3A, 3B etc. which are used every time again for different geographical indication)
- the year the renewal has been carried out
- the component renewed:
    - switch rail including accompanying stock rail divide by direction e.g.:
        - left leading on a left turnout
    - screw hole renovation
    - welding on the frog
    - grinding
    - checkup

A full description of the database including some results on the data mining can be found in annex 6.

5.1.4 Loads on the SBB network database

From the complete SBB railway network the loads are obtained for the period 1947 – 2005, as registered in the database Carico. The information is divided in:
- passenger train loads
- freight train loads
- service train loads (maintenance and renewal trains, rescue trains)

The database is divided for uniform sections of lines where the loads are the same. This is mainly from node to node, which can be a station, an intersection, a cross-over or a junction. For the largest amount of S&C it I not possible to determine precise enough in which direction how much trains went (expressed in tonnage).
The results are expressed in annual cumulative tonnages.

5.1.5 Other databases

Other databases include information on:
- the exact location relative to the Swiss national geometrical grid
- the main direction a track is used (allowing to state if a turnout is mainly used in facing or trailing direction

5.2 *The search fro a model: generalities*

The methodology used for this study consists of several steps:
- Database setup and control (establishing the combined database);
- Composing the degradation model;
- Testing the model;

Using the model to obtain the objectives as formulated in chapter 1 of this document.

Two models have been tested on their functionality:
1. Markov-chain model;
2. a reliability model.

They are described below in §5.3 and §5.4 respectively.

5.3 The Markov-chain approach to the degradation and wear description

One of the methodologies used in maintenance optimisation research especially related to LCC is the Markov-chain approach. It consists of a description of the current state, which changes in a future state by a probabilistic step. The only information taken into account is the current state, which contains all the information. So the probabilistic step is independent from how the current state has been reached.

A test sample was made to test this method. Since there is no data on the actual state available, the age in years or the cumulative tonnage of a switch or crossing or its component, had to be taken as a "current state" which would evolve in a future state. The only two states possible are that a switch is OK or too old and has to be replaced. As the initial state is chosen the distribution for the cumulative tonnages. For the transition a model was necessary, but the only transition parameter that could be obtained, related to the tonnage, was the average replacement rate.

The result of it was that for each specific subset of switch or crossing (per switch type and switch angle and curved or not etc.; cf. to §5.1.1 for all the different parameters) an analysis was necessary to get a reliable outcome. This result is the same however as an analysis of all the different average replacement ages with the same high degree of specific result. Beside this, the effect of different measurable parameters (switch angle, soil quality etc.) in relationship to eachother would not be visible. In other words, the effect of different parameters cannot be stated, also because the sample size would get too small.

It is noted that for a Markov Decision Programming, reward values were necessary, closely related to e.g. reliability figures. But since these are also not available, this method is abandoned.

It was here that is was mentioned to use a model which could consider the parameters separately and together: a reliability model with simple single parameter or multi-parameter linear regression.

5.4 The reliability model for the degradation and wear description

Degradation or wear results almost always in a certain failure, or at least in a need for maintenance or renewal[45]. The most interesting parameter from the point of view of infrastructure management is the probability that this failure will occur at a certain time, i.e. that the reliability of a certain S&C or S&C component cannot be guaranteed. Thus a way to describe the effect of degradation or wear is with help of a probabilistic model in the form of a reliability analysis[46] as described by Omori (2003), Hausman & Woutersen (2004) or Meier-Hirmer (2007).

In this case the reliability is the probability that a switch, a crossing, or an S&C components functions properly longer than some specified time (equation 9).

---

[45] Because of some non-redundant properties, components have to be replaced before they actually "fail", because failure would lead to an immediate and dangerous incident a derailment. Failure means in this case therefore also: does not pass the functionality test.
[46] also known as *duration analysis* or *duration models* in economics or sociology, or *survival models* or *survival analysis* in biological studies

$$R(t) = P(S > t) \qquad (9)$$

where

- $R(t)$ reliability function expressed in t
- P probability
- t time or tonnage[47]
- S time or tonnage when failure occurs and a maintenance or renewal action is necessary (i.e. in this case the actual maintenance or renewal action)

During its lifetime, a switch, crossing or S&C-component can be maintained or replaced at moments S, for which is valid:

$$(S_n)_{n\geq 1} \qquad (10)$$

and that S is ascending in time (e.g. a second replacement cannot take place before the first replacement) – cf. figure 3.

Figure 45 Renewal or maintenance points, from Meier-Hirmer (2007)

The reliability function $R(t)$ is presumed to approach zero when age t increases, which is true: at a certain moment an S&C-component or a complete switch or crossing will be replaced.

The available database provides for several S&C-components and complete switches or crossings their age t during replacement, in as well time as tonnage. With this, the lifetime (the time that an S&C-component, a switch or crossing does not fail) distribution function F can be defined as the complement of the reliability function:

$$F(t) = P(S \leq t) = 1 - R(t) \qquad (11)$$

The derivative of F is known as f and is known as the event density - the failure rate per time (or tonnage, cf. the footnote below):

$$f(t) = \frac{d}{dt} F(t) \qquad (12)$$

When taking into account the fact that this counts for the maintenance or replacement actions starting with S_1 (cf. equation 2 and figure 3), than the probability that a maintenance or renewal action takes place, can also be written as:

---

[47] An interesting feature of railways is that maintenance and renewal needs can be described in time as well as in tonnage; a total tonnage of 20.000 tons (comparable with the passage of 35 Intercity trains) can pass during one day or during two days. But in the last situation, the wear or degradation of the track *in time* will go approximately twice as slow. Recalculated in tonnage however, this problem will be overcome – this is the reason that t (and S) in equation 1 can be expressed in time as well as in tonnage, but in case of time, the tonnage has to come back in the formula.

$$P(S_1 \leq t) = F(t) = \int_0^t f(s)ds \tag{13}$$

For the second time that a similar maintenance or renewal action has to be carried out/the second time that the same failure occurs, this can be described as:

$$P(S_2 \leq t) = \int_0^t \int_0^\infty f(s-u)duds \tag{14}$$

This probability may be calculated for all the S_n, which means that for all maintenance and renewal actions as mentioned in Table 14 and Table 15.

5.4.1 Single parameter analysis

The results can be presented as single parameter or a 1-to-1 analysis of the parameters. This means that single relationships can be presented based on the presumption that different parameters are not correlated. It will lead to a probability function of which the effects of different influential parameters are obtained separately.

By choosing a basic distribution which fits best, a reliability model can be derived.

At the end this will result in a formula in the form of a probability function that a certain switch, crossings, or a certain switch or crossing component will need replacement or maintenance action:

$$P(S_n) = \varepsilon + \sum_a^z \alpha \cdot F_{a...z}(t)_n \tag{15}$$

where

P	probability
S_n	time or tonnage expressed in t when failure occurs and a maintenance or renewal action is necessary for the n-th time
a	constant
$a...z$	(summarized) for every different lifetime distribution of a parameter having an effect
$F_{a...z}(t)_n$	distribution functions a to z for a certain parameter, for the n-th time that this component is replaced
ε	rest term

In case multiple parameters have a proven effect, it is convenient to continue immediately to the multi-parameter analysis.

5.4.2 Multi-parameter analyses

As stated in the first part of this chapter, some parameters have a correlation, although their relationship is not quantified. There are several ways to arrive at a quantification of these correlations:

- correlation diagrams (scatter plots) or schedules;
- multi-parameter automated regression using robust regression methods in case of outliers.
- Pearson product-moment coefficient in case of a linear correlation;
- Spearman rank correlation coefficient (a simple case of "Pearson", in case there are no tied ranks).

With these tests correlations can be quantified and a critical analysis can be made on the necessity of all variables.

5.4.3 Parametrical analysis

At this stage it is also possible to state if a parametrical analysis is possible or not. If this is not possible (because of parameters being dependant) the derived empirical lifetime distribution functions F can be used (cf. *single parameter analysis* in combination with results of the Kaplan-Meier estimator) to derive results for different subgroups of the parameters (for example the parameters soil quality in combination with direction, speed etc.).

A parametrical model however will use a stepwise regression analysis, to determine for each parameter (e.g. soil quality, switch type, speed of the trains) the effect of it on the life expectancy of a switch, a crossing, or one of its components, taking into account different distribution models (Weibull, exponential etc) and calculating their variables (e.g. λ and k for Weibull, λ for the exponential distribution). The result will then be a ready-to-use finished model which can easily be incorporated in an LCC model.

This way of testing can be understood also as an enormous vector [y] for the loads and matrix [X] for the influential parameters. It is the following equation which has to be resolved:

$$\vec{y} = \vec{X} \cdot \vec{c} \tag{16}$$

in which

$$\vec{y} = \begin{bmatrix} y_1 \\ y_2 \\ \vdots \\ y_n \end{bmatrix}$$ are the cumulative tonnages

$$\vec{X} = \begin{bmatrix} 1 & x_{\alpha 1} & x_{\beta 1} & x_{\chi 1} & \cdots \\ 1 & x_{\alpha 2} & x_{\beta 2} & x_{\chi 2} & \\ 1 & \vdots & \vdots & \vdots & \\ \vdots & & & & \end{bmatrix}$$ are the parameters influencing the maximum tonnage a switch or crossings (component) can bear before a maintenance or renewal action is necessary.

$$\vec{c} = \begin{bmatrix} \varepsilon \\ \alpha \\ \beta \\ \vdots \end{bmatrix}$$ are the coefficients to be determined.

It does however need a matrix which is sufficiently cleaned, to derive rationalised models.

5.5 Testing of the degradation model and introduction of a maintenance policy parameter

Independent if the model turns out to be a parametrical one, or a non parametrical one, the model can be tested in the same way. That is by feeding the situation in 1997 (the first complete year available in the database) and simulating the same traffic load (amount of trains) during the years 1997-2007. The difference between the results from the simulation and the actually carried out work form an input for improvement of the model and to introduce the parameter "maintenance policy" in the model. Maintenance policy contains the complete process of decision making (strategic, tactical and operational); parameters other than the technical ones.

An example is the strict maintenance and renewal policy on both the Lötschberg and Gotthard route: the 2 major routes for passing the Alps. These two routes are heavily loaded, during day and night with both freight trains and passenger trains. The fact that timetable-disrupting incidents on these lines can cause capacity problems from the inlands of Italy until the middle of Germany, due to halted trains, make it necessary to have a high reliability and availability of the infrastructure. This leads often to earlier replacements of components than strictly necessary. This can be found in the data by relative early replacements, compared to similar loaded tracks, with less high reliability and availability demands. These routes are also used by piggy-back trains (carrying trucks, the so-called *Rollende Landstrasse*) using wagons with extremely small wheels. Due to the higher derailment risk, sensitive parts of the tracks with regard to these wheels (diamond crossings/frogs of switches and crossings, switch tongues) are therefore earlier replaced.

An extra "maintenance policy" parameter would therefore be necessary to introduce. Another way to retrieve this parameter is through interviews with experts on this matter.

The following test of the model will be to simulate the maintenance and renewal needs for the period 2007-2017 or even further.

5.6 Integration in a Life Cycle Cost model

The results from the previous sections can be used in the maintenance and renewal management tool which focuses on Life Cycle Cost reduction. With the help of a cost database annual budgets can be calculated. Improvements are possible by introducing efficiency in time (by combining works in time) as well as place (by combining works on local, regional and national scale). For the 2007-2017 simulation different strategies can be compared: the effects on the cost of different traffic loads with no optimisation, optimisation in time or optimisation in place.

It would be further interesting to see what would happen if the choice of switches and crossings would be optimized in case it is found that certain types show less wear when subjected to the same load, even when they are more expensive to install.

A last optimisation that can be made is the analysis of interchangeable works. An example of this is the inexpensive but less sure of success repair-welding of a piece of metal on a frog vs. replacing the frog completely.

In this way the objectives as stated in chapter 1 are reached.

5.7 Final remarks on the methodology

The required databases should be regarded with prudence. There are a lot of standard turnouts according to this dataset replaced far before their designed life expectancy in as well tons as years.

This might have easy explanations like noise (faults) in the database, integral renewal (a turnout is renewed together with the track, even when not necessary, but for example because the track was built with another rail profile), infrastructure improvements or abandonment of infrastructure and thus the switch connecting this abandoned infrastructure to the main line.

An option for this problem would be to introduce *truncation* on the dataset, which means that switches and crossings with a life time less than a certain threshold are treated separately and not together with the big dataset on which the model is based. Already a first start of this is made by taking only the main tracks into regard and not secondary lines – although most of the switches and crossings can be found in the latter. Further truncation or separation would be possible if exceptional maintenance and renewal policies have been identified.

6. Modelling of degradation & restoration process

This chapter describes how the models are established with the modelisation as described in the previous chapter. To derive the models different databases had to be combined, below is described how this has been done and which filters have been applied. Also a list of presumptions regarding the analyses is provided.

6.1 Data handling and database combination

The different databases available are described in the previous chapter. Several links had to be established which are not easily done (no unique parameter to link tables):
1. the link between the switches and crossing databases and the load over the network
   o after step one, the attribution of cumulative loads from the initial installation of a switch or crossing until a maintenance work or renewal work is carried out or the switch has to be replaced;
2. the combination of the switches and crossings and their respective soil quality from the soil quality database;
3. the combination of the switches and crossings and their geometrical place in respect to the national Swiss grid.

The steps are described below.

6.1.1 Attributing the load

The load database is composed of registration per line section. A section can be a station, a node, a switch a crossing for which the length equals the length of the switch or crossing, a stretch of a couple of kilometres of plain track between two switches or crossings. Of these sections it was indicated which ones are switches and crossings. With the help of the node name (in German *Betriebspunkt*) and the local switch number including (e.g. 306) and in some cases an extra sign (e.g. 306B and 306A for connecting crossovers) the link could be made with the unique switch number, and thus all other S&C properties.

The load is summarized annually, but can be easily converted to a value per month or day by dividing by respectively 12 or 365. However, the date of certain actions (a maintenance or renewal work) is expressed most of the time in years only. Therefore the calculation of the cumulative tonnage was:

$$T_S = 0.5 \cdot T_n + T_{n+1} + T_{n+2} + \ldots + T_{m-1} + 0.5 \cdot T_m$$

where
T cumulative tonnage
index S the maintenance or renewal action
index n year of construction/installation
index m year in which S took place

Although this simplification introduces a non existing accuracy in the results, it is closer to reality than the division per year: during winter time, less maintenance and renewals are carried out, leading to more maintenance and renewal during the summertime, which is approximately the middle of the year.

6.1.2 Attributing the soil quality

The soil quality is expressed in the DfA by one of the three numbers 1, 2 and 3, in which 1 means high quality soil and 3 for low quality. Low quality in this case means: with bad or unequally deflecting soil. It can also indicate a poor drainage. This results into a high chance of geometrical degradation.

The soil quality is indicated per section, however most of the time the indication for a section with a switch or crossing was not available. Experts confirmed that it are not the track constructional properties and conditions of sub ballast, ballast, sleepers and rails, determining the soil quality, but the existing soil. Therefore, the assumption was mad that if in the section before and after the switch or crossing, the soil quality parameter has the same value, that this is also valid for the respective switch or crossing.

6.1.3 The position of the S&C in Switzerland

An analysis per GIS system of the available data showed where the subjects of this study are located:

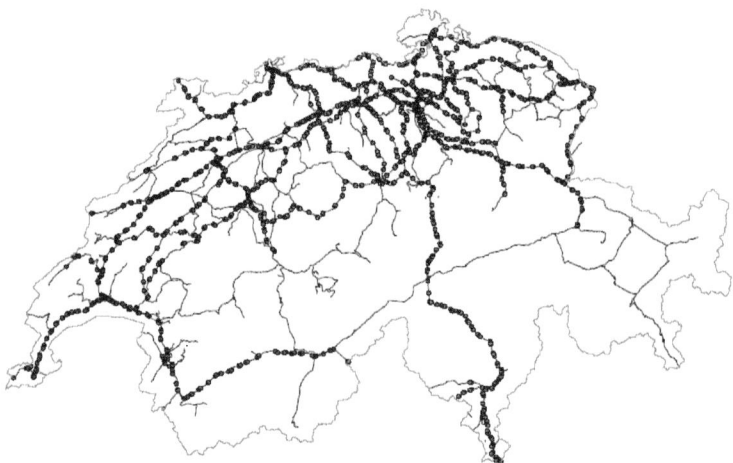

Figure 46 Map of Switzerland indicating the place of the switches and crossings subject of this study[48]

This map confirms that all switches and crossings on the SBB network are included and non-SBB S&C excluded. This counts e.g. for the Lötschbergline, which is managed and maintained by the BLS and several regional lines.

6.2 Applied filters

There are several filters applied on the data before hand. The database on which this analysis was carried out, WEITABL1, contained at the start 15.326

The filters are applied in the order as presented.

---

[48] Source of the railway lines on this map: TeleAtlas, borders of Switzerland: Swiss Federal Statistical Office

1. Restricted to the track gauge of 1435mm. The number of unknown track gauges and other than 1435mm track gauge totalled 837. These are mainly S&C on the Brünigbahn (rest in database: 14.489)
2. Restriction to S&C where SBB is 100% owner. These are mainly switches and crossings not on main tracks, which supply access to private tracks, e.g. those towards a factory. This done because switches and crossings not 100% property of SBB might not be subject to comparable maintenance policies. (12.475 left)
3. Restriction to S&C where SBB is responsible for 100% of the maintenance and renewal (12.357 left)
4. Separate analysis can be made for S&C on slab track (25 pieces), but due to the low sample size, they are not of importance for this study and therefore note taken into account. All other support types than normal sleepers (e.g. steel profiles used on bridges or special constructions to support the switch or crossing) are also excluded also because of their small numbers (12.274 left). A separate analysis of some turnouts on slab track show significant high renewal ages, for some components even twice as long as switches on normal soil.

With this cleaned database the analyses are carried out.

6.3 Presumptions

A couple of presumptions are made before starting the analyses.

I. *All switches & crossings for which the complete renewal is analyzed were on wooden sleepers or exceptionally on metal sleepers.*

From the former switch or crossing only the amount of years it was in service are exactly known. The date of installation of the current switch or crossing equals the date of removal of the previous. The installation year of the former switch or crossing can then be calculated by subtracting the amount of years in service from the installation year of the current switch. It is only since recent that new main line switches and crossings have been equipped with concrete sleepers, at least since a shorter period than their expected life time.

II. *All switches and crossings for which the complete renewal is analyzed had the same geometry (angle and curve radius) as the current switch.*

From the renewed switch or crossing nothing exactly is known

III. *The maintenance and renewal policy is the same:*
  a. *per track category*
  b. *over the last 30 years*
  c. *per switch type and component type*

The result from this presumption is that maintenance and renewal actions were always executed at the same wear condition level, unless it was a timely planned maintenance activity.

IV. *The replacements and repairs have been carried out for a non-significant (or even negligible part) for reasons of accidents, i.e. all replacements or maintenance actions carried out can be regarded as necessary because of degradation, tear, wear, plastic deformation, or because they were timely planned.*

6.4 Complete switch or crossing renewal

One of the most expensive works to be carried out is the renewal of a complete switch or crossing. It is therefore also that this work is done only in case of necessity. However, many reasons can make it a necessity, and these are not only ear related. Sometimes a switch or crossing is replaced because of a speed increase, making it necessary that the technology is improved to allow passage with higher speeds or higher loads.

The decision to renew a switch or crossing strongly depends on the maintenance policy as shown in Rivier, Putallaz & Zwanenburg (2005). And since a high amount of money is involved it also strongly depends on the available budget, which can vary during the years.

The renewal rate for standard turnouts therefore differs per year and so does the average age of the renewed switches or crossing. It should also differ with the parameters for which it is thought they influence the degradation and wear.

6.4.1 Replacement ages of complete switches and crossings during the years

In the three figures below, the replacement age for simple turnouts in track categories HG1, HG2 and HG3 is provided. It shows the different

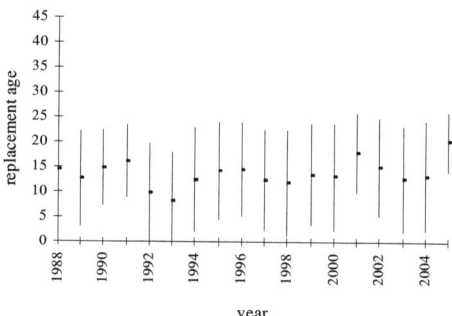

Figure 47 Average replacement age for simple turnouts in HG1 tracks (1988-2005).

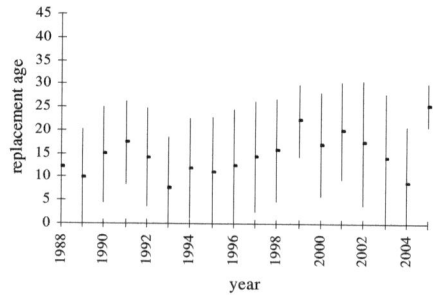

Figure 48 Average replacement age for simple turnouts in HG2 tracks (1988-2005).

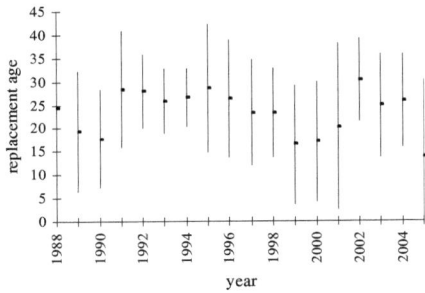

Figure 49 Average replacement age for simple turnouts in HG3 tracks (1988-2005).

The figure below shows for double diamond crossings with slips in track category HG1 the average replacement age.

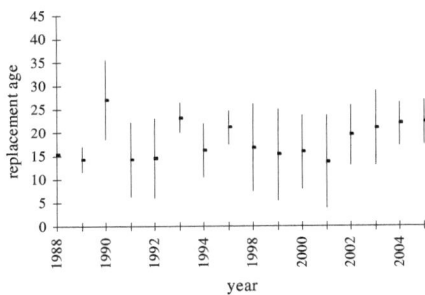

Figure 50 Average replacement age for double diamond crossing with slips in HG1 tracks (1988-2005).

The average values of what is displayed in the four graphs above, can be found in Table 16.

Table 16 Average replacement age and standard deviations for different S&C types (1988-2005)[DfA, SBB]

Track category	S&C type	Average replacement age (years)	Standard deviation (years)
HG1	simple turnout	13.7	2.7
HG2	simple turnout	14.9	4.7
HG3	simple turnout	23.8	4.3
HG1	double diamond crossing	18.2	4.0

The sample size for other track categories and S&C types is too small to perform any statistical analysis.

Figure 47 to Figure 50 and Table 16 allow the conclusion that there is a relationship between the average load and the S&C life expectation: the higher the load, the shorter the expected lifetime. However, the relationship is weak, because a turnout in HG3 is exposed to a load, at least 2.5 to up to 3 times higher. The relationship weakness is therefore mainly due to large standard deviation in the results and the too rough division into 3 track categories, although a much finer division in track loads is possible. Beside it, this also points out that there are apparently other factors affecting the S&C life expectation.

What also can be seen is an increase in the average duration an S&C stays in the track. Possible reasons for it might be that maintenance or component renewal is being preferred over a complete renewal (change of maintenance policy – which is presumed not to be the case and confirmed by SBB experts it is not), and the longer expected lifetime of new generation S&C (improved rolling geometry causing less wear, better materials e.g. hard wood and concrete instead of timber sleepers, heavier rail profiles, better support etc.).

6.5 Complete switch or crossing replacement modelling

6.5.1 Data analysis

From 9297 switches and crossings in the filtered database it could be retrieved how long the previous switch (before the last complete replacement) was in place (VORHER in the database WEITABL1). There is no information stored on these switches and crossings, but some assumptions can be made:
- There is a high probability that these switches or crossings are of the same type (standard turnout, symmetrical turnout, diamond crossing etc) and geometrical lay-out (frog angle, curve radius) as the current switches and crossings:
    - when a switch is replaced during a reconstruction and thus not at its original position, often the new one receives another number and is not remarked as an follow-up of its predecessor;
    - most of the time only new technology is introduced (concrete sleepers replace wooden sleepers), the geometry stayed the same;
- Because only recently concrete sleepers were introduced, it is probable that the replaced switches and crossings were on wooden sleepers and for a minor part on steel sleepers.

This dataset is combined with the database for loads. The combination with tonnage was not possible for all S&C replacements. This is mainly due to the fact that the exact location in the network could not be automatically found. This delivers the results per switch or crossing type and per track category (situation: 2005).

Table 17 Number of replaced switches & crossings per type, of which the life in years and tonnage is known [DfA, SBB]

S&C type	Number of S&C Life in years known	Number of S&C Life in tonnage known
EW	7574	2270
SW	171	-
GD	57	1
EKW	49	12
DKW	993	155
DW	258	1
MS	178	3
other	17	-
Total	9297	2442

Table 18 Number of replaced switches & crossings per track category, of which the life in years and tonnage is known [DfA, SBB]

Track category	Number of S&C lifetime in years known	Number of S&C lifetime in tonnage known
HG1	2334	1747
HG2	649	425
HG3	445	201
NG1	331	35
NG2	1671	32
NG3	3463	2
other	404	-
Total	9297	2442

From the two tables above, it can be concluded that for the most common switches and crossings (i.e. the standard turnout (EW) and the diamond crossings with double slips (DKW)) and in the most common track category (HG1, 2 et 3) an analysis seems statistically relevant, because of sample size. Those will be presented in the paragraphs below.

The following parameters are analysed:
- Time and tonnage
- Switch angle or curve radius
- Curved or not
- % of freight trains from tonnage database
- Soil quality
- Main train direction
- Speed

And the results are divided in single-parameter analysis and multiple parameter analysis.

6.5.2 Results for a standard turnout (EW)

6.5.2.1 Single parameter analysis for EW

Time and tonnage
For a standard turnout, the number of replaced turnouts per track category shows that the biggest amount of renewal takes place in track category HG1. Only the amount of replacements in HG1, HG2 and HG3 is regarded as relevant.

Table 19 Number of replaced standard turnout per track category [DfA, SBB]

Track category	replaced standard turnouts
HG1	1619
HG2	400
HG3	191
NG1	30
NG2	28
NG3	1

The renewal "age" of a standard turnout (EW) can be expressed in years and tonnage. A comparison of these two ways is visible in Figure 51.

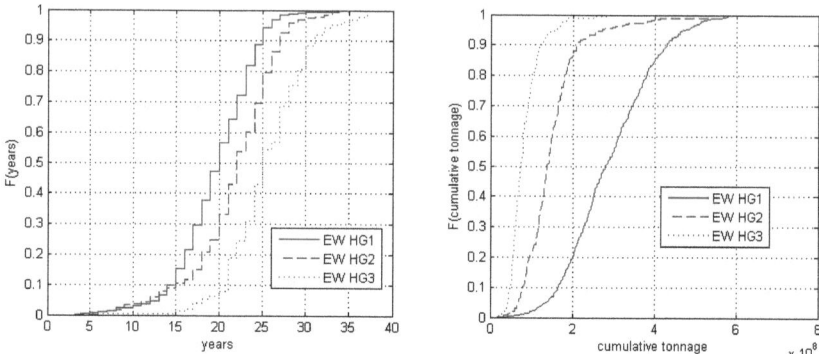

Figure 51 Cumulative lifetime distribution functions of standard turnout replacements in years (left) and tonnage (right) [DfA, SBB]

From the two graphs in Figure 51 can be concluded that the standard turnouts subjected to a higher daily tonnage (HG1) in general stay a shorter time in track. However, the right graph shows, that these turnouts will endure a higher load during their life in track. Since HG1 represents a more than 2 times higher daily load than HG3, it should be expected, that turnouts in HG3 track should stay at least 3 times longer in track – which is not the case. There might be several reasons for this, on of them are the possible environmental effects, which might have a bigger effect than load on S&C subjected to a light load only, leading to a renewal necessity because of e.g. wooden sleepers degradation away.

Experts confirmed that it is the condition of a standard turnout determining its renewal needs (exception: replacements because of track reconstruction works and as a result of derailments), which means that it is the load determining this state for HG1 and that it might be the environment for HG2 and 3. Otherwise, the fact that for HG2 90% and HG3 99% is replaced before 200 million tons of cumulative load, while at the same time, similar switches are subjected to twice the amount of cumulative load in HG1, can hardly be explained.

Therefore the rest of the analyses, which take the load as most important parameter, do not take into account HG2 and HG3. This strategy is confirmed to be valid for plain track by Veit (2005) and Gaudry & Quinet (2003).

Another interesting effect visible in the figure above is that 20% of the standard turnouts in HG1 are replaced before 200 million tons of cumulative loads and before being in place 16 years. These do not necessarily have to be the same turnouts. A cross-check shows the relationship between cumulative loads and time-in-place for standard in HG1 of 1619 turnouts.

Table 20 Combination of cumulative tonnage when replaced and amount of years in place [DfA, SBB]

		cumulative tonnage before replacement					
		0-100 million tons	100-200 million tons	200-300 million tons	300-400 million tons	400-500 million tons	500-600 million tons
time in place [years]	1-6	13	2	0	0	0	0
	7-12	9	23	12	0	0	0
	13-18	7	61	125	193	37	0
	19-24	2	145	280	241	139	37
	25-30	2	56	154	44	9	17
	>30	0	0	4	4	2	1
	Total	33	287	575	482	187	55

Although HG1 switches and crossings are subjected to a load of 30.000 cumulative tons per day nowadays (approximately 11 million tons annually) and at least daily 25.000 tons (9.2million tons) before the SBB directives were updated a couple of years ago, it is interesting to see there are some standard turnouts replaced with only a small cumulative tonnage (bottom left part of Table 1) and at the same time a high amount of years in place.

This can have several reasons, but one of the most important can be a change in the use of the track: what was before perhaps a seldomly used track, was later used more heavily when the traffic on the network got denser which is the case on several spots in Switzerland during the investigated period. This is for example true for the line Bern-Lausanne, which was a HG2 line before and is now HG1 (10-25 or 30 kilotons per day vs. more than 25/30ktons per day). In the renewal modelling for this study this effect is taken into account by letting the age count more if it is getting higher, even when a purely-tonnage based analyses would not show a need for replacement. Taking Figure 51 into account, this would for example mean that every switch will be replaced after 35 years in track. The degradation modelling itself will from now on be only expressed in tonnage.

Switch angle or curve radius

The switch angle or, better, the frog angle, is expected to have an effect on the switch lifetime expectation.

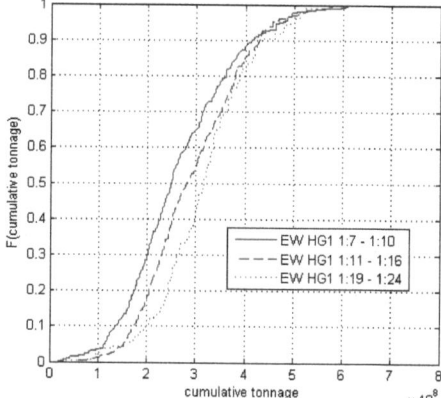

Figure 52 Cumulative lifetime distribution functions of standard turnout replacements related to the turnout angle [DfA, SBB]

It can be seen in the figure above that standard turnouts with a small angle (1:19 to 1:24) last longer in the track (expressed in tonnage) at least for the first part of the figure. There might be several reasons for this:

- A small angle switch is used on locations where in both directions trains can use higher speeds. These are places where apparently a speed restriction, as necessary with big angle turnouts, would cause to much delay or capacity constraints. This is the case at important junctions and not so much at station entrances (low train speeds). On these important junctions, the turnouts tend to be used in both directions (straight and curved) in a significant amount. This leads to more equal loading, on the contrary to the one-sided loads of big angle turnouts, which are mainly used in one direction.
- Because of their placement at crucial junctions and thus important places in the network, they might be better maintained and thus less subject of a component-wear-inducing bad geometrical state.
- The fact that small angel turnouts are significant more expensive than big angle, might make their renewal only later economically feasible.
- Turnouts with a small angle might be passed at high speeds also in the curved direction. Only this high speed provokes the maximum amount of lateral acceleration or jerk. Every speed slower than that causes a lateral acceleration or jerk smaller than the maximum allowed. The chance that a train will pass with a smaller speed than allowed is bigger in case of a smaller angle switch 80km/h turnout, than with a 40km/h big angle turnout.

A similar analysis is possible for a parameter closely related to switch angle: the switch radius.

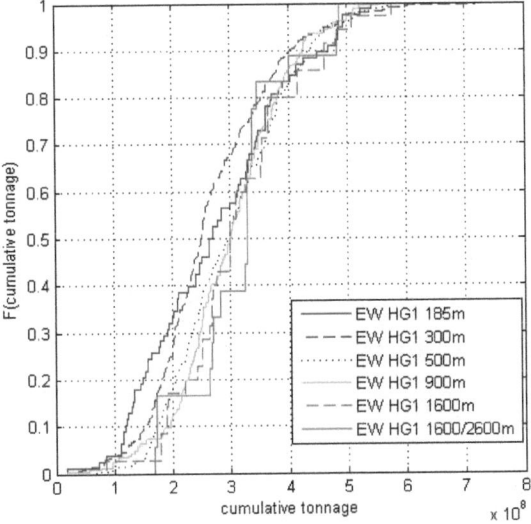

Figure 53 Cumulative lifetime distribution functions of standard turnout replacements related to the turnout radius [DfA, SBB]

Also here it is visible, that the large radius (related to a small angle) last longer in the track. It can thus be stated that switch angle and/or radius are important.

Curved or not
A general opinion among railway experts is that curved switches (figure 7) are more expensive than straight ones (figure 8), because they are subjected to higher loads and more difficult to position correctly. This might lead to a shorter expected life time.

Figure 54 On the left a curved turnout (both directions are in a left turn), on the right a top view on a normal turnout with one straight direction and one curved direction

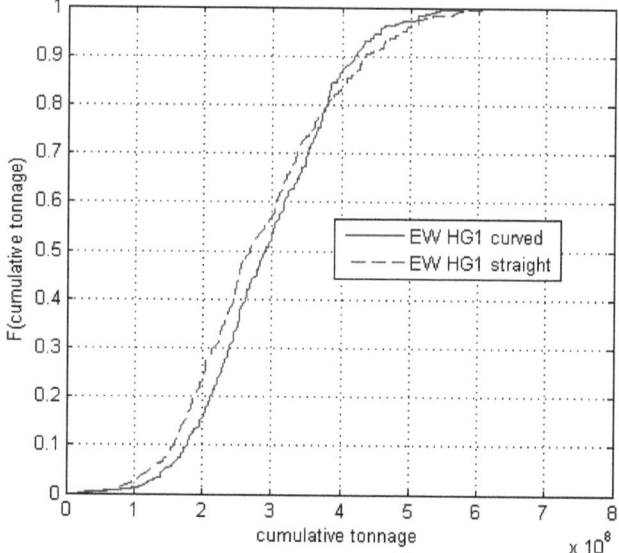

Figure 55 Cumulative lifetime distribution functions of curved and straight standard turnouts [DfA, SBB]

From Figure 55 can be concluded that, at least for this dataset, no significant difference could be found for the distribution of expected life duration of a curved or straight turnout subjected to heavy loads. Therefore this is not taken into account for the modelling of complete turnout renewals.

% of freight trains from tonnage database

There are no direct measurements for the actual axle loads. But following about thi subject is known:
- An electrical locomotive of any type on the Swiss railway network, has a maximum axle load of 21 tons;
- Passenger train wagons have axle loads of approximately 10 to 15 tons;
- EMU have axle loads ranging from 10-20 tons;
- The maximum axle load for freight wagons is 22,5 tons;
- In all directions on the Swiss railway network, trains run both loaded and unloaded: over the Gotthard cargo is transported both towards Italy as well as from Italy to the north. It is therefore presumed that loads are equally divided per direction.

Based on the experience with and studies on the effects of high(er) axle loads as explained in chapter 3 and 4 of this report, it can be presumed that the high axle loads of freight trains have a measurable effect on the wear of the tracks and thus S&C. The effect of the percentage of freight train tonnage as part of the total tonnage, on the turnout life expectancy (in cumulative tonnage) is expressed in Figure 56

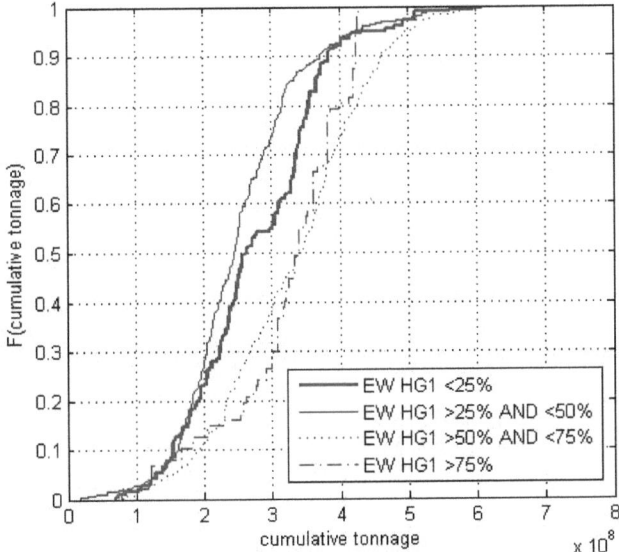

Figure 56 Cumulative lifetime distribution functions related to the percentage of freight trains [DfA, SBB]

Concluded from figure 6 is that turnouts subjected to a lot of freight trains (expressed as a percentage of the total load), are staying longer in track than those who have a more significant share of passenger trains. This is on the contrary to what was expected. Standard turnouts with less than 50% of freight train tonnage are replaced at already a much earlier stage.

An expert at SBB reflected this by stating that tracks with only freight trains are not only rare, but also less likely to be subjected to heavy maintenance actions, even on HG1, these tracks are allowed to degrade more, with temporary speed restrictions imposed. For the modelling this is a significant parameter.

Soil quality
From 709 standard turnouts, the current soil quality could be derived from the available database. It is a rough indicator, provided by the track inspection and management crew, based on long time experience. There are only 3 values: 1, 2 and 3, in which 1 indicates the highest quality of soil.

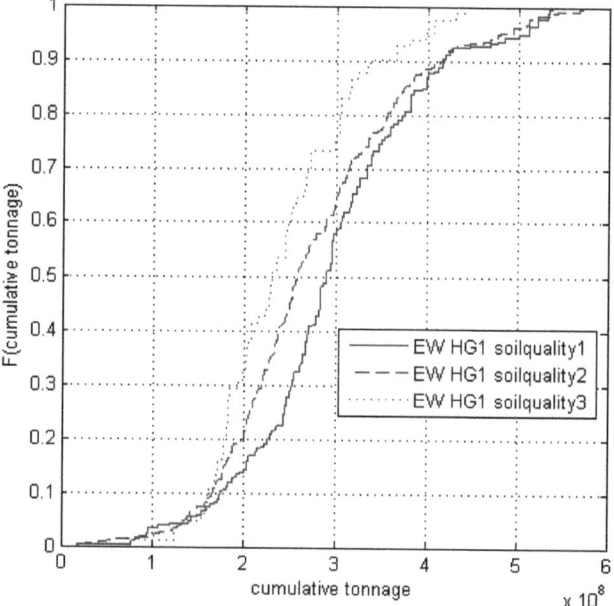

Figure 57 Cumulative lifetime distribution functions related to the soil quality [DfA, SBB]

It can be concluded that there is a significant difference in life-expectancy from a turnout when placed on a good or a bad soil. Bad soil as indicated by quality three, shows a significant shorter life expectancy than turnouts on good soil. This is a significant parameter for the analyses.

Main train direction
It is said that the direction that trains mainly take on a turnout influences its wear. There are 4 directions a standard turnout can be used:
1. facing straight direction
2. facing curved direction
3. trailing straight direction
4. trailing curved direction

Unfortunately, the database only allows an analysis for trailing and facing direction. For a standard turnout, the results are displayed in Figure 58.

6 - Modelling of the degradation and restoration process

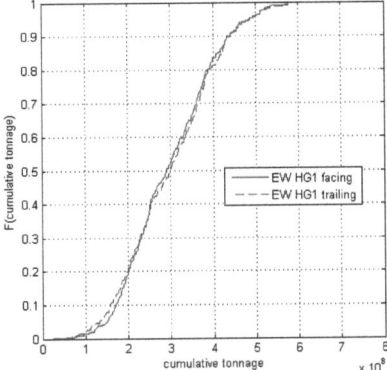

Figure 58 Cumulative lifetime distribution functions related to the main direction a standard turnout is used [DfA, SBB]

Speed
From a limited amount of turnouts the speed in through direction could be determined. The highest speeds are taken for the calculation (R135, for trains with a brake percentage >135%). Freight trains are mostly limited to 80 or 100 km/h.

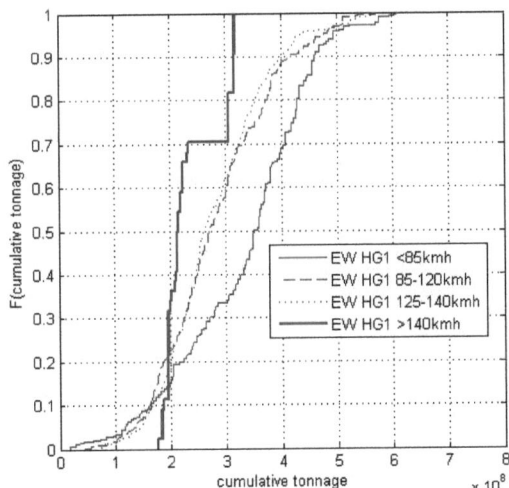

Figure 59 Cumulative lifetime distribution functions related to the percentage of freight trains [DfA, SBB]

Visible here is that the life distribution for turnouts used with speeds in the straight direction of 85-140 km/h does not differ much. However, albeit the small sample size, the turnouts used with speeds >140 km/h are subjected to a significantly lower life expectancy. An SBB expert mentioned, that this might be related to program renewals, to upgrade tracks to higher speeds. Switches and crossings as for that reason replaced by more modern types (on concrete sleepers), which are better able to resist the high loads of trains running at higher speeds. Due to the way the database is composed, the registered speed is the current train speed and not the speed with which the earlier switch was used. However, the difference between the group of turnouts subjected to train speeds >85 km/h and those of 85-140km/h is also significant. This clearly shows that turnouts close to yards and railway stations, where speed is limited, do have a longer life. The reason for this is probably the reduced dynamic loads, which is related to trains speed as described in chapter 3 and 4.

<u>Conclusion on the single parameter analyses of the lifetime distribution of standard turnout in HG1-tracks</u>

From the sections above can be concluded that following parameters have an effect:
- tonnage for heavily loaded standard turnouts, time for standard turnouts with a low load
- switch angle or radius
- percentage of freight trains
- speed

No effect on the life distribution function have:
- if a turnout is positioned in a curve or not;
- if a turnout is mainly used facing or trailing.

Since more than one parameters has an effect, immediate continuation to the multi-parameter analysis is valid.

6.5.2.2 Multi-parameter analysis for EW

There are several ways to derive a multi parameter analysis. One is to group the results from the single parameter test. The percentage of freight trains shows two groups (<50% and ≥50%), similar for the switch angle or curvature, which can be divided into 3 groups (185-300 m radius, 500-900 meter radius, >1600 meter radius), the soil quality only knows 3 categories and also the speed knows three distinct results in the cumulative lifetime distribution (<85km/h, 85-240 km/h and >140 km/h).

This would result in 2x3x3x3 = 54 lifetime distribution functions. However, for a high number of functions, the amount of standard turnouts for which all data is available might turn out to be limited. That's why it is necessary to turn to multi-parameter regression, using regressions over all the data.

The above mentioned 4 parameters (tonnage, switch angle, percentage of freight trains and speed have been analyzed for their correlation. If a linear correlation would exist it would show by values close to -1 or 1. This is not the case as the table below shows.

Table 21 Correlation of parameters influencing standard turnout renewals

	%freight trains	switch angle	soil quality	speed
% freight trains	1.0000	0.1174	-0.1798	-0.0995
switch angle	0.1174	1.0000	-0.1033	0.0628
soil quality	-0.1798	-0.1033	1.0000	0.0493
speed	-0.0995	0.0628	0.0493	1.0000

The matrix X is composed of the values:
- x_1 representing the percentage of freight trains;
- x_2 the switch angle;
- x_3 the soil quality;
- x_4 representing the speed.

After passing the T-test, normal, linear regression and robust regression are then executed on the data. Unfortunately only from 248 standard turnouts all data of the 4 parameters was available. Since the data is prone to outliers, a robust linear regression is made as an alternative. This uses a uses an iteratively reweighed least squares with a bi-square weighting function as algorithm (standard in *Matlab 2007b*).

Table 22 Multi-parameter regression analysis (values to be multiplied by 10^7 to retrieve tonnage equivalents)

	%freight trains [0-100]	switch angle [1-12][49]	soil quality [1,2,3]	speed [40-200]
normal linear regression result	0.2091	1.4529	0.6792	0.0598
minimum weight(95% range)	0.1590	0.9603	-0.9215	0.0221
maximum weight (95% range)	0.2592	1.9455	2.2799	0.0976
robust linear regression result (rest =21.98)	0.1980	0.7850	-1.2020	-0.0610

A big difference between the normal linear regression and robust linear regression is mainly an indication for outliers in the results: their influence is reduced in the robust linear regression. The robust linear regression shows results more in line with the results of the single parameters: higher values of soil quality and speed lead to reduced turnout life expectancy. Therefore, they should have a negative value, but they only have that in the robust analysis. It is the result of the robust linear regression which is taken into account for the modelisation.

6.5.3 Results for a diamond crossing with double slips (DKW)

The much smaller sample of renewed diamond crossings with double slips (the second most common switch and crossings on the SBB network), makes that not all track categories are equally presented, as Table 23 shows.

Table 23 Number of replaced diamond crossings with double slips per track category [DfA, SBB]

Track category	Replaced diamond crossings with double slips
HG1	117
HG2	22
HG3	9
NG1	5
NG2	2

Due to this limited sample size the analyses distinguishing track category, curved or not or the soil quality are not possible. The analyses on main train direction is also not useful,, since a diamond crossings with double slips is symmetrical, on the contrary to a standard turnout.
This means that remaining for analyses are:
- Time and tonnage
- Switch angle or curve radius
- % of freight trains from tonnage database

---

[49] In which 1 stays for 1:7, 2 for 1:8, 3 for 1:9, 4 for 1:10, 5 for 1:11, 6 for 1:12 7 for 1:14, 8 for 1:16, 9 for 1:19,. 10 for 1:21.5, 11 for 1:24 and 12 for 1:25.

- Speed

6.5.3.1 Single parameter analysis for DKW

Time and tonnage
Just like for standard turnouts in the previous chapter, a comparison can be made between the time a diamond crossing has been in place and the cumulative tonnage it was subjected to during that time.

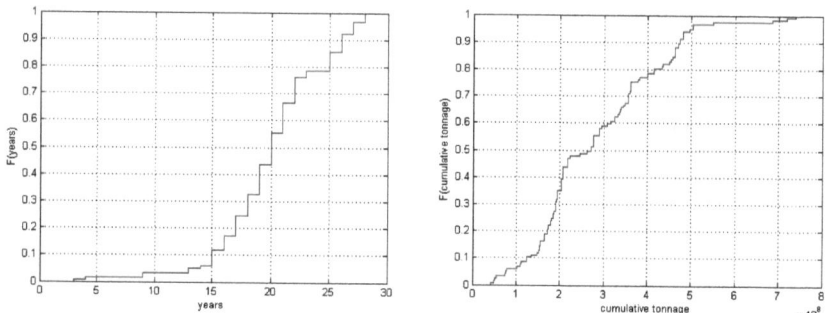

Figure 60 Cumulative lifetime distribution functions of diamond crossings with double slips replacements in years and tonnage [DfA, SBB]

What can be seen from the figures is that serious replacement here only starts after 15 years in place. However also here in tonnage a much more variance in results can be found, with replacement linear divided between 100million and 500 million tons of cumulative load.

Switch angle or curve radius
There are only 2 types of diamond crossings with double slips available in the dataset in significant volumes: one with a 1:8-angle and a 160 m radius, the other with a 1:9-angle and a 185 meter radius.

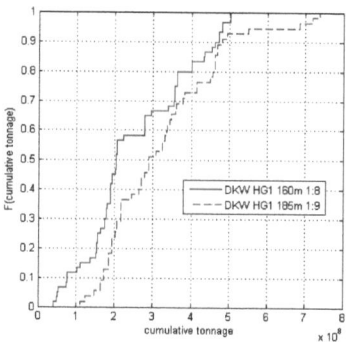

Figure 61 Cumulative lifetime distribution functions related to the switch radius/frog angle [DfA, SBB]

From this it can be concluded that the differences between the two are significant and that they should be taken into account when developing the renewal model for diamond crossings.

% of freight trains from tonnage database
Here can be found the same result as for a standard turnout: a high percentage of freight trains apparently also indicates another maintenance and renewal philosophy. Although the sample size is relatively small (visible through the corners in the lines in the graph), the difference is significant.

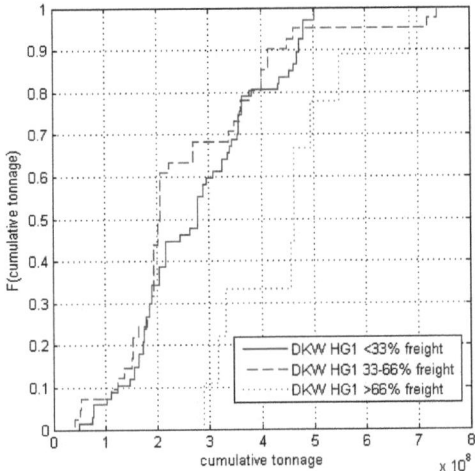

Figure 62 Cumulative lifetime distribution functions related to the percentage of freight trains [DfA, SBB]

Speed
A diamond crossing with double slips is almost never allowed to pass with a higher speed than 125 km/h, due to high amount of unguided openings and the resulting jumps from train wheels from one rolling surface to the other.
The speed and the resulting differences in dynamic load could have an effect on the expected time the DKW will be in place.

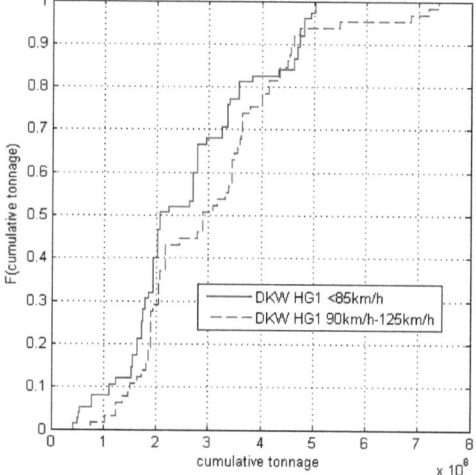

Figure 63 Cumulative lifetime distribution functions related to the percentage of freight trains [DfA, SBB]

Although the two speed groups have some differences, the shape of the lines is not as far apart as with a standard turnout, so this parameter is not influencing the life of a switch much.

<u>Conclusion on the single parameter analyses of the lifetime distribution of diamond crossing with double slips in HG1-tracks</u>
From the sections above can be concluded that following parameters have an effect:
- tonnage for heavily loaded standard turnouts, time for standard turnouts with a low load
- switch angle or radius
- percentage of freight trains

These will be taken to the multi-parameter analysis. The speed has no effect on the life distribution function.

6.5.3.2 Multi-parameter analysis for DKW

The same multi-parameter analysis is used as for standard turnouts before in this appendix.

Table 24 Multi-parameter regression analysis (values to be multiplied by 10^7 to retrieve tonnage equivalents)

	%freight trains [0-100]	switch angle [2,3][50]
normal linear regression result	0.1633	9.0465
minimum weight(95% range)	-0.040	6.4650
maximum weight (95% range)	0.3310	11.6280
robust linear regression result (rest =10.24)	0.1300	5.2630

The result from the robust linear regression will be taken to the DKW renewal model.

---

[50] In which 2 means 1:8, 3 for 1:9.

6.6 Synthesis

The result from the database analyses as presented in above shows that according to the available data, the renewal need of standard turnouts (EW) in track category HG1 is influenced by:
- the percentage of freight trains;
- the switch angle;
- the soil quality;
- the speed.

No influence was measured for the fact if the standard turnout is curved or not, and the direction.

Beside the standard turnout (EW), only the sample size of the diamond crossing with double slips (DKW) is big enough to carry out a significant analysis. From the sample size from this crossing type, it could only be derived that the percentage of freight trains and the frog angle have any significant effect.

As presented in chapter 5.4.2, the problem can be described in the form of an equation with two vectors and a matrix $\vec{y} = \vec{X} \cdot \vec{c}$.

The values of the matrix are now defined and summarized in the table below.

Table 25 Result of the parametrical test on complete switches and crossings

		Parameters of influence						
Term	rest term ε	%freight trains [0-100] x_1	frog angle [0-12][51] x_2	Crved or not [1-2] x_3	Soil quality [1, 2, 3] x_4	Direction facing trailing [1, 2] x_5	Speed [35-200km/h] x_6	multiplication factor
complete switch or crossing								
EW HG1	21.89	0.1980	0.7850	-	-1.2020	-	-0.0610	10^7
DKW HG1	10.24	0.1300	5.2630[52]	-	-	-	-	10^7

The results for the standard turnout (EW) in HG1 are consistent:
- on tracks with a high percentage (85% or higher) of freight trains, the standard turnouts tend to stay longer in place. According to an SBB expert that was because it is easy to impose a speed restriction on these tracks;
- the life expectancy tends to decrease with the switch angle (=increase with the curve radius): a smaller angle leads to a longer life (a larger curve leads to a longer life);
- the bad soil quality (indicated with a 3, on the contrary to a good soil quality which is indicated with a 1 indeed has a negative effect on the expected lifetime;
- the speed has a negative linear relationship.

---

[51] In which 1 stays for 1:7, 2 for 1:8, 3 for 1:9, 4 for 1:10, 5 for 1:11, 6 for 1:12 7 for 1:14, 8 for 1:16, 9 for 1:19,. 10 for 1:21.5, 11 for 1:24 and 12 for 1:25.
[52] Only the values 2 and 3 are used

The results for the diamond crossing with double slips (DKW) in HG1 are also consistent:
- the frog angle is largely determining in the wear rate: the 1:9 will resist much higher loads than a 1:8;
- the percentage of freight trains also has an effect but much smaller than the switch angle.

It is therefore proposed to continue with these numbers to the model test in the next chapter.

7. Model testing

This chapter presents the results from the testing of the model with the results from Table 25 in the previous chapter.

7.1 Determination of renewal needs for standard turnouts

The test as described in chapter five, consist of several phases.
1. start with the situation from beginning 1997, obtained by subtracting from the database from 2005 what happened in 2004, 2003, 2002, 2001, 2000, 1999, 1998 and 1997 etc.
2. define the state of the switches and crossings in 1997
    a. determine the age
    b. calculate the cumulative loads to which a switch or crossing was subjected in the period before that age

It is presumed that factors like soil quality, switch angle or radius or speed have not changed since then.

The model will calculate for every year if a threshold related to the expected life time of the switch or crossing.

Following results were the first one to be obtained for the 1600 standard turnouts in EW. From those, for only 613 all parameter data could be obtained (%freight trains, frog angle etc.), with as most reducing parameter, the soil quality, which is for very little amount of switches sufficiently available.

Table 26 Results for the modelling

Year	1997	1998	1999	2000	2001	2002	2003	2004
Real replacements	13	37	43	26	42	36	21	26
Model replacements	90	5	7	10	6	14	10	9
similarity	0%	0%	0%	0%	0%	0%	0%	0%

Several problems can be indicated:
1. In 1997 is clearly a starting problem visible in which a large amount of switches according to the model has already passed the threshold and should be replaced. This is typical for imposing the results of a multi-parametrical where this is not taken into account.
2. The amount afterward of the turnouts which according to the model should be replaced, is far lower that the actual amount carried out.
3. The turnouts of which it is indicated that they should be replaced are in not a single case the turnouts which are in reality really replaced.

With regard to point 3, it can be extended to the fact that also not a single replacement as indicated as necessary has happened in another year: Per year as well as for the total period, there is no similarity.

However, the model parameters are consistent as shown in the previous chapter and there are turnouts passing their replacement threshold according to the model.

7.2 Reflections on the obtained model

There are several reasons thinkable, as explained also on other places in this report for the fact that the final model has only a reduced validity.
- an amount of switches is replaced because of infrastructural renewal, without really a clear link with the actual replacement need because of wear or degradation;
- an amount of switches is replaced because of infrastructure upgrades to withstand better the ever increasing loads on the Swiss Railway network and to introduce more reliable, modern and easier to maintain switches;
- other options related to the maintenance policy.

The question remains how these factors can influence the renewal parameters in such a way that the effort to make a model based on wear or degradation (through an indirect analyses of the renewal-data) does not provide a satisfying result at the end. Another question is why the above mentioned reasons happen anyway.

Part of the answer to these question is provided in a benchmark between Swiss and foreign maintenance standards. It is known from Rivier (1994) and Rivier, Putallaz and Zwanenburg (2005) that the infrastructure components (track, signals, catenary, switches and crossings, but often also tunnels and bridges) of the SBB network are relatively young compared to neighbouring countries: they have a low average age, which is the result from low replacement ages. In other words: the Swiss renew their track components early. This counts for main track components (in straights or curves we only find ballast, sleepers and rails), but also for switches and crossings. This strategy is the result of a maintenance policy in which 2/3 of the budget is used for renewals and 1/3 for maintenance. The budget itself is however not that high compared to other countries; the amount of money spent per track kilometre or per train-kilometres per track kilometre is even modest. But it must be stated that no separate analyses for switches and crossings is provided in Rivier, Putallaz and Zwanenburg (2005).

For plain track, as stated in the introduction chapter of this report, maintenance and renewal have been highly standardized, mechanized and is now being well planned with decision support tools. In the mean time, these tools lack the knowledge to also forecast maintenance and renewal needs for switches and crossings. In the total budget for track maintenance and renewal, there is therefore a big part for which spending is very well planned, i.e. for plain track, while for switches and crossings it might still be a decision by people, without a proper analyses if the money is well spent and if the money is really spent on the right time.

This can therefore lead to much more money being spent on switches and crossings than what would really be an economical and technical optimum, but it is invisible in the total sum due to the highly rationalized way in which plain track maintenance and renewal is planned and executed.

The result for this study is that maintenance and renewal decisions or not based on rationality, but perhaps even only on budget restrictions of which nobody knows if these limits are right or not.

To this challenge can be added another one, i.e. that a large part of the renewals have taken place because of track upgrades to support heavier or longer trains and/or in case of general renewals in which complete stations were renewed, unrelated to the state (wear and degradation level) of all track components (plain track, switches and crossings).

These points are objects of discussion with SBB experts. They should remain points of discussion, because these results show that there are some discrepancies between a sound maintenance and renewal policy. Such a maintenance and renewal policy should result in a working model and the proper study results.

7.3 Synthesis

The model is performing bad because of pollution of the data with which it is filled. The sample size (which are all switches and crossings on the SBB railway network, and everything that happened to these switches and crossings during the period 1997-2005), is big enough to state something on the parameters that influence the maintenance or renewal need. But for more detailed analyses, e.g. to derive a model which describes the influence of all data to derive maintenance and renewal needs, the database needs to be cleaned.

This cleaning involves first the introduction of a distinction if a switch or crossing has been replaced because of wear or degradation reasons (or even end of economic life), or that it is part of an investment program to upgrade or improve infrastructure. In the last case, these switches and crossings, should not be taken into account.

In a second case, all switches and crossings which have been replaced before they are at their end of life because of other reasons (e.g. the two switches from a crossover might be replaced simultaneously, but only one was worn, the other wasn't, or because the renewal crew and equipment were during a certain moment present in a certain area), should also be taken out of the analyses.

The first point can quite easily be achieved and perhaps executed on the currently available database, as long as information on investments during the period 1997-2005 can be retrieved. The second one is much more difficult to obtain, since it is related to another problem: the inexistence of actual degradation data which states something on the actual state of a certain switch or crossing at a certain moment. Due to this it is also impossible to derive the actual state when replacement took place. To achieve this, a much more rational way of inspection should be introduced.

With the database cleaned for these things, at least proper effects of influential parameters (e.g. cumulative loads, axle loads, soil quality) can be obtained and both maintenance and renewal can be optimized (and not only renewal as is the case now). This would also allow to rationalise maintenance and renewal and to apply decision support systems to create network wide optimisations, which are too big to oversee as one person.

Also the optimisation with respect to plain track renewals is then possible, i.e. that they can take place together.

8. Conclusion

8.1 General conclusion

It is interesting to see that in Switzerland the maintenance and renewal of switches and crossing is highly standardized and mechanized. Abroad, the SBB is mentioned for its standardisation and proper quality of deliverance. However, at the same time, it is hardly possible to capture the need for this maintenance or renewal in a model as proposed and presented in this report. Although the derived parameters and influences seem logic and defendable – in a sense that they confirm common knowledge – a working model is still subject of discussion and future research, but the framework is set in this study.

8.2 The parameters that cause wear and degradation

The different parameters which are supposed to influence the switch or crossing degradation have sometimes clearly shown their effect. These parameters are for standard turnouts (EW):
1. the actual train load: more train loads mean more wear;
2. the soil quality: lower soil quality leads to a significantly reduced life expectancy of a switch;
3. the frog angle: a smaller angle switch generally has a longer life;
4. the percentage of freight trains, or more precise, the axle loads: higher axle loads lead to more wear;
5. the speed: the higher the trains speed on the through track, the more wear and degradation can be observed, resulting in a shorter life expectancy;

Of no significant influence on the life expectancy were the fact if a standard turnout is curved or not – on the contrary to what many believe in foreign countries – and the main direction the trains were using a turnout (facing or trailing only).

For diamond crossing with double slips (DKW), these influences are
1. the actual train load: more train loads mean more wear;
2. the frog angle: a smaller angle switch generally has a longer life;
3. the percentage of freight trains, or more precise, the axle loads: higher axle loads lead to more wear;

The soil quality, if it is a curved type or not and the speed did not have a significant correlation with the life expectancy of a diamond crossing with double slips.

8.3 The results of the multi-parameter model

The correlations from §8.2 can however not be quantified with the used multi-parameter linear model.

The fact that the model does not work can be attributed to various causes. One of the most important of them is that due to investments in infrastructure improvements exactly during the complete period of analysis, a large amount of switches and crossings have been upgraded (replaced by a better type, e.g. a turnout on wooden sleepers replaced by one on concrete sleepers or a big angle turnout by one with a smaller angle for capacity reasons) without the necessary need from a wear or degradation part of view. The switches were simply not at the end of their life. It might very well be that it was just the right moment to replace because both replacement crew and crane were there. In general this should not have a big effect on the results, if it would have happened on a smaller scale- The investments for Rail2000 were however so widespread over the SBB network, that their effect on the actual maintenance and renewal numbers was probably significant. This has a very large effect on the composition of a model, as presented in this report, and composing a model has thus become an extreme challenge, sinceit is now shown that it is extremely difficult to do.

8.4 The added value of this thesis

Although the results have not lead to a working model, some interesting insights have been retrieved:

- the data shows a strong relationship between subsoil quality and the expected maximum age of a standard turnout, an extra reason to do the utmost to improve soil quality when installing or replacing a turnout;
- the analysis has shown that smaller angle switches will last longer, which means that although they cost more to build and install, their extra cost may be (partially) compensated due to a longer life expectancy. It must be stated that smaller angle switches also allow higher speeds and if in that case a train can pass at higher speeds in the curved direction, that it brings also external benefits through energy saving and wear of the train brakes. The other consequence might be related to capacity: if a turnout can be used in the curved direction with a higher speed, a train does not have to break that much, or can accelerate to a higher speed, allowing a quick passage and liberation of the occupied track;
- with regard to the speed in the through direction (straight), three categories can be distinguished for the expected lifetime: <85 km/h , 85-140 km/h and >140 km/h, with the last one, the most long lasting;
- the expected life of both S&C and components increase significantly if the elasticity is more homogeneous as for example on better soils or a track on a concrete slab.

Beside this, there are lessons which can be taken in mind for future changes in inspection, design, maintenance and renewal challenges described in the recommendations (next chapter)

9. Recommendations

The main recommendation is that this study should continue but based on the following improvements:
- improved input of the infrastructure management on the switches and crosses which are chosen to compose the model, i.e. select only the switches and crossings which are replaced because they of wear and degradation reasons;
- a better insight and registration of the reason for replacement and renewal, so that replacements because of infrastructure improvements can be left out of the model;
- to obtain more insight in the annual decision processes other than available in databases.

Taking a sample from another period, geographical area or railway network in which less investments in infrastructure improvements have been done, would probably upgrade the models reliability. It might be an idea to analyse all renewals in a year with significantly less renewals than normal.

For this study a statistical approach is chosen. Due to a lack of financing a more engineering-like approach (using tests in laboratories, or on a real switch during several years) was not possible. Laboratory studies or theoretical approaches exist and should be coupled to this or similar studies. This study for instance, might give an indication, that the soil quality –which is seldom easy to vary in laboratory or real tests– is of much bigger influence on the life expectation than material properties of e.g. the rails, train wheels or the rail pads (rubber or cork plate between sleeper and the rail), or the geometrical design of railhead and train wheel.

It is quite astonishing to see that after 6 years, the recommendations covering all aspect of switches and crossings as mentioned in Jovanović *et al.* (2002) are still valid. Only small progression has been made since then.

9.1 Inspection of switches & crossings

The inspection of S&C in Europe still largely takes place on foot. Although sophisticated devices exist to measure with high precision and high accuracy the state of plain track <u>and</u> switches & crossings, only a couple of trains nowadays exist, able to take over this task, e.g. the *Videoschouwtrein* [video inspection train] in the Netherlands. Although this is a special self-propelling train or wagon, it will not be long until the task of this train is combined in measuring trains which measure at the same time geometrical degradation and component wear phenomena. The latter exist already for multiple years.

A sophisticated measuring train has multiple advantages:
- High precision and objective measuring in all weather conditions;
- Measurement under load;
- Speed (a measuring train can measure at high speeds due to faster computation power and faster and more accurate video-data processing);
- Therefore cheap (one measuring train is able to cover an enormous amount of tracks annually if it is continually used);
- Safety for personnel (no need to go on the tracks anymore for inspection)

It can only be recommended that the development and the application of these trains are not limited, because regulations prescribe a certain type of measurement (often imprecise and out-of-date) as obligatory. The development of new measuring techniques should give a chance to inspect switches and crossings more often, but with less effort and in a way that information on it can be stored digitally.

Another development should be the introduction of a more condition-based inspection frequency. After original teething problems, switches and crossings tend not to show very much wear the first period of its life. Inspection frequency, especially those on foot can be downgraded, especially when a couple of tasks are taken over by the inspection trains as mentioned before.

9.2 Design of switches & crossings

The design of switches and crossings has not changed much since the beginning of the railways. This might mean that current switch designs are simply perfect. But visible at the same time are the cost involved for building, inspecting and maintaining them. One of the most common problems is the fact that they malfunction more often than plain track, simply because they have more things which can malfunction. There are two directions visible how the down-time due to these problems is reduced: more simple robust switches, especially with regard to the switch actuator/motor – the main source of malfunctioning, and to apply continuous (on-line) condition monitoring to detect and identify problems in early stages. The use of the information from condition monitoring systems is limited to the operational phase. It should however also be possible to obtain more information from it e.g. long term trends or results regarding better design.

A switch or crossing is, as described multiple times in this report, subjected to high dynamic loads, since the switch or crossing it self introduces a heterogeneity in the support for the passing train at different levels: a different elasticity of the ballast bed, longer sleepers, lighter and heavier rails which are attached in another was to the sleepers etc. Although recent design changes, e.g. the FAKOP system have dealt with some problems regarding the dynamic loads when a train passes through a switch, much more effort should be put into it. Even taking into account not only acceleration, force or jerk analyses, but also of an analysis on energy level. The total energy dissipated through the rail/wheel contact should be as little as possible: a new way of regarding the design of railways, but also the maintenance and renewal. If they are optimized for the lowest amount of energy, a track condition might be derived which can be regarded as a technical-economic optimum.

9.3 Maintenance of switches & crossings

The maintenance of railway tracks and thus S&C is a heavy task with a high amount of heavy components and necessary heavy equipment. Luckily, the introduction of sophisticated machines and maintenance trains providing protection for track workers have improved working conditions, but still the amount of maintenance need is high.
This is also due to the high forces on small areas to which (components of) S&C are subjected, is a reduction of these forces, by whatever means will automatically lead to a reduction of maintenance. This thus might also be the result of introducing lighter trains.

9.4 Renewal of switches & crossings

In Switzerland, the renewal of S&C is a highly mechanized and standardized procedure. It can be done during the night, with only one track of a double track out of service. Switches and crossing arrive in plug-and-play status in several components which can be put together during a night. In other countries this is about to be introduced. But often is forgotten that this way of working in Switzerland is the result of year long experience with this method and the fact that from service level, the crews are forced to work like this. Also all necessary equipment (among it large cranes) and skilled personnel is available. It is also the complete chain that has been optimized; from the

design phase (which takes into account the way in which the switch will be transported in big parts and lifted and where for that reason extra support is already designed in before production), via transport on special wagons and placement on a prepared ballast bed.
This way of working is now introduced in several countries, but some are hesitant. It does require investments and changes in design and logistics. However, the switches will be more accurately placed (which is an advantage) because they will not be constructed at site, but for the biggest part in a factory.

Another improvement might be obtained by better soil improvements, even when replacing an existing switch.
Since it is also shown that switches with a smaller angle cause less wear, replacement of existing switches and crossings by the same type (but better technology, e.g. concrete sleepers in stead of wooden sleepers and better rails), might be reconsidered taking into account the secondary benefits of turnouts with smaller angles. Performance improvements might there be lower LCC than purely on the basis of investment cost is derived and less usage of energy and wear of train brakes if trains can pass at higher speeds.

9.5 Research of switches & crossings

Research on switches and crossings is limited to the following fields:
- studies regarding small improvements of existing designs, carried out in-house at S&C manufacturers;
- scientific research on mechanic wear phenomena.

However, only a small amount of studies is really carried out on an optimisation of all above mentioned recommendations. Research can focus more on optimal maintenance and renewal planning, based on optimisations not currently used in railways.

Optimise the inspection frequency by using the degradation and wear predictions as stated in this study, might also be a research subject of interest.

Optimise the inspection itself by using more automated equipment, i.e. handheld for inspections on foot and using video-measuring trains with automatic analysis of the gathered pictures. The gathered information of all inspection should be stored to understand the wear phenomena better.
This should also result in a proper database allowing the optimisation S&C design.

10. References and literature

10.1 General

Andersson, C. & T. Dahlberg (1998) "Wheel/rail impacts at a railway turnout crossing" *IMechE Journal of Rail and Rapid Transit*, Vol 212, no F2, pp 123-134

Berg, G. & H. Henker (1978) "Weichen" Reihe Eisenbahnbau, Transpress Berlin, Germany

Bonaventura, C. et al. (2006) "Optimizing Vehicle Dynamics Through A Switch While Maintaining Existing Switch Lead Length" *Proceedings of the Joint Rail Conference, 4-6 April 2006*, Atlanta, USA

Clayton, P. (1996) "Tribological aspects of wheel-rail contact: a review of recent experimental research" *Wear 191*, pp. 170-183

Dang Van, K & M.H. Maitournam (2002) "On some recent trends in modelling of contact fatigue and wear in rail" *Wear 253*, pp. 219-227

Dekker, R. (1995) "On the use of operations research models for maintenance decision making", *Microelectronics Reliability*, Vol. 35, Nos 9-10, pp.1321-1331

Esveld, C. et al. (2001) "Modern Railway Track" 2nd edition, MRT Productions, Zaltbommel, the Netherlands

Ferreira, L. & M. Murray (1997) "Modelling rail track deterioration and maintenance: current practices and future needs" *Transport Reviews*, Vol 17 No 3, pp. 207-221

Fischer, F.D., Oberaigner, E.R. et al. (2005) "The impact of a wheel on a crossing" *ZEVrail* 129 – 8 , pp.336-345

Fischer F.D., W. Daves (2008) "The impact of a wheel on a crossing" *ZEVrail* 132, pp.187-190

Gaudry, M. & E. Quinet (2003) "Rail track wear-and-tear costs by traffic class in France"
Université de Montreal, Publication AJD-66

Hausman, J. & T. Woutersen (2004) "A semi-parametric duration model with heterogeneity and time-varying regressors – draft paper" *MIT and John Hopkins University*, November 2004

Hettler, A. (1984) "Bleibende Setzungen des Schotteroberbaus" *ETR Eisenbahntecgische Rundschau 33 Vol. 11*, pp.847-853

Hiensch, M., Larsson, P.-O., Nilsson, O., Levy, D., Kapoor, A., Franklin, et al. (2005) "Two-material rail development: field test results regarding rolling contact fatigue and squeal behaviour" *Wear 258 (7-8)*, pp. 964-972.

Higgins, A., Ferreira, L. & M. Lake (1999) "Scheduling rail track maintenance to minimise overall delays" *Transportation and Traffic Theory*, pp.779-796

Ishida, M. et al. (2001a) "Gauge Face Wear Caused with Vehicle/Track Interaction"; *Proceedings of the World Conference on Railway Research 2001*

Ishida, M. et al. (2001b) "Track settlement measurements and dynamic prediction model based on settlement laws" *Proceedings of the World Conference on Railway Research 2001*

Jovanović, S. & W.-J. Zwanenburg (2002) "Switches and Crossings Management System: EcoSwitch – Feasibility Study" *ERRI Project Report D251/RP1* European Rail Research Institute, Utrecht, The Netherlands

Kassa, E. & G. Johansson (2006) "Simulation of train-turnout interaction and plastic deformation of rail profiles" *Vehicle System Dynamics*, Vol.44, Issue S1, pp. 349-359

Kassa, E., Andersson, C. & J.C.O. Nielsen (2006) "Simulation of dynamic interaction between train and railway turnout" *Vehicle System Dynamics*, Vol. 44, Issue 3, pp. 247-258

Koller, G. (2008) "Kunstholz für den Gleisbau" *Der Eisenbahningenieur 04/2008* p.24-27 VDEI, DVV Media Group, Hamburg Germany

Larsson, D. (2002) "Verification of DeCoTrack – A prediction Model for Degradation Cost of Track", *Research Report, Division of Operation and Maintenance Engineering*, Luleå University of Technology, Sweden

Larsson, D. (2004) "A Study of the Track Degradation Process Related to Change in Railway Traffic", *Licentiate thesis*, Luleå University of Technology, Sweden

Lewis, R. & U. Olofsson (2004) "Mapping rail wear regimes and transitions" *Wear 257*, pp.721–729

Lichtberger, B. (2005) "Track Compendium – Formation, Permanent Way, Maintenance, Economics" Eurailpress Tetzlaff-Hestra Verlag, Hamburg germany

López Pita, A. (2005) "Optimizing the quality of the wheel/rail system, cost-efficiency, financing. Very high-speed track design and cost optimization in Spain" *Railway Technical Review* 3/2005

Lüdeking, R. & P. Götz (2007) ",,Herzstückauframpung": Verbesserung von Weichen und Kreuzungsherzstücken" *EI-Eisenbahningenieur*, Oktober 2007, pp.58-59

Mace, S., et al. (1996) "Effects of wheel-rail contact geometry on wheel set steering forces" *Wear 191*, pp.204–209

Marschnig, S. & P. Veit (2006) "LCC optimised permanent way strategies for track and turnout components" *ZEVrail Glasers Annalen* 130 11-12, pp.500-508

Mauer, L. (1995a) "An interactive Track-Train Dynamic Model for Calculation of Track Error Growth" *Vehicle Systems Dynamics Supp 24*, pp. 209-221

Mauer, T. (1995b) "Hochgeschwindigkeit auf Weichen und Schienenauszügen – Erfahrungen, Folgerungen, Entwicklungen" *ETR-Eisenbahntechnische Rundschau (44) No. 6*, pp. 440-445

Meier-Hirmer, C. (2007) "Modèles et techniques probabilistes pour l'optimisation des stratégies de maintenance. Application au domaine ferroviaire" *PhD thesis* Université de Marne-la-Vallée, France

Meng, H.C & K.C. Ludema (1995) "Wear models and predictive equations: their form and content" *Wear 181* pp.443–457

Nicolai R.P. & R. Dekker (2007) "A review of multi-component maintenance models" *Risk, Reliability and Societal Safety – Aven & Vinnem (eds)*, pp.289-296

Olofsson, U. & T. Telliskivi (2003) "Wear, plastic deformation and friction of two rail steels—a full-scale test and a laboratory study" *Wear 254*, pp.80–93

Omori, Y. (2003) "Discrete duration model having autoregressive random effects with application to Japanese diffusion index" *Journal of the Japanese Statistical Society* Vol. 33, No1, pp.1-22

ORE C116 rp10 (1981) "Study of optimum rail inclination and gauge related to wheel profiles adapted to wear" Utrecht, the Netherlands

ORE D117, rp2, 5, 7, 12, 18, 20, 26, 29 "Optimum Adaptation of the conventional Track to Future Traffic", Utrecht, the Netherlands.

ORE D141 rp5 (1982) "Study of the technical and economical consequences of increasing the axle load from 20 to 22.5t", Utrecht, the Netherlands

Prud'Homme, A. (1970) "La Voie" *Revue Générale des Chemins de Fer*, No. 1, pp. 56-72

Putallaz, Y. (2007) "Gestion stratégique de la maintenance et de la capacité d'un réseau ferrée" *PhD thesis* 3852 EPFL, Lausanne, Switzerland

SBB (2008) "Richtpreisliste 2008. Weiche, Stellwerk- und Signalmaterial, Fertigungsteile, Schwellen, Schienen, Befestigungen und Kabelverteiler" Infrastruktur Verkauf Bahntechnik, Hägendorf, Switzerland

Rivier, R.E. & C. Hofmann (1994) "Politique et coûts prévisionnels de maintenance du réseau national des voies ferrées", SNCF, Paris, France

Rivier, R.E. (1999) "Life Cycle Cost Analysis and EcoTrack: Economical Track", *Presentation UIC Headquarters*, Paris

Rivier, R.E. (2000) "Superstructure de la voie ferrée" Lecture notes *"Gestion de la maintenance des réseaux ferroviaires"* EPFL – LITEP, Lausanne

Rivier, R.E. (2002) "Maintenance d'ouvrages et d'infrastructures" Lecture notes *"Gestion de la maintenance des réseaux ferroviaires"* EPFL – LITEP, Lausanne

Rivier, R.E., Putallaz, Y., Zwanenburg, W.-J et al. (2005), "Audit sur l'état du réseau ferrée national français, synthèse" LITEP – EPFL, Lausanne

Sato, Y. (1995) "Japanese studies on deterioration of ballasted track"; Vehicle System Dynamics Vol 24 Supp., pp. 197-208

Sawley, K.J. (2001) "Wheel/Rail Profile Maintenance" *Proceedings of the Wolrd Conference on Railway Research*, Köln, Germany

Scheffers, S.A. (2007) "Degradatie in beeld" *MSc thesis* Technische Universiteit Eindhoven & Movares, Netherlands

Schmitt, L. (2006) "Recent SNCF research on ballasted high speed track fatigue behaviour" *Workshop Proceedings Track for High-speed railways* 12-13 Octobre Porto, Portugal

Schwab, C.A. & L. Mauer (1989) "An interactive Track/Train Dynamics Model for Investigating the Limits in high-speed tracks" *Proceedings of the 11th IASVD Symposium, Kingston U.S.A Supplement to vehicle System Dynamics, Vol. 18*

Selig, E.T. et al. (1981) "A Theory for Track Maintenance Life Prediction", *U.S. Department of Transportation*, Rept. DOT/RSPA/DPB-50/81/25

Shenton, M.J. (1975) "Deformation of Railway Ballast under repeated loading conditions", *Proceedings of a Symposium held at Princeton University*, Pergamon Press

Steenbergen, M.J.M.M. & C. Esveld (2006) "Rail weld geometry and assessment concepts" *Proceedings of the Institution of Mechanical Engineers, Part F: Journal of Rail and Rapid Transit* 220 pp. 257–271

Steenbergen, M.J.M.M. (2008a) "Quantification of dynamic wheel–rail contact forces at short rail irregularities and application to measured rail welds", *Journal of Sound and Vibration* 312 p.606-629

Steenbergen, M.J.M.M. (2008b) "Wheel-rail interaction at short-wave irregularities", *PhD thesis* Delft University of Technology, Netherlands

Thomas, L. (1986), "A survey of maintenance and replacement models for maintainability and reliability of multi-item systems" *Reliability Engineering 16*, pp.297-309

UIC leaflet 714.R "Classification des voies des lignes au point de vue de la mainteannec de la voie" *leaflet 714.R*, 1.1.1989

Ulrich, D. & M. Luke (2001) "Simulating rolling contact fatigue and wear on a wheel/rail simulation test rig"; *Proceedings of the World Conference on Railway Research 2001*

Vos, E. (2003) "Studie naar wegvriendelijke vrachtwagenbanden" Ministerie van Verkeer en Waterstaat – Directoraat-Generaal Rijkswaterstaat, the Netherlands

Veit, P. (2005) "Outsourcing of Track Maintenance" *Presentation EFRTC General Assembly*, slide 8-11

Veit, P. (2006) "Wirtschaftlich optimaler Abstand von Überleitstellen" *ZEVrail* Glasers Annalen 130 – 3, pp.116-121

Wang, H. (2002) "A survey of maintenance policies of deteriorating systems" *European Journal of Operational Research* 139, pp.469-489

Warmerdam, D.L.M. (2005) "RAM(S) aspecten voor het ontwerpen van wissels" [RAM(S) aspects for the design of switches & crossings], *MSc thesis*, TU Delft & Holland Railconsult, Utrecht, the Netherlands

William, J.A. (1999) "Wear modelling: analytical, computational and mapping: a continuum mechanics approach" *Wear 225-229*, pp.1–17

Zhang, Y.-J., Murray, M. & L. Ferreira (1997) "Railway track performance models: degradation of track structures" *Road & Transport Research*, Vol 6 No 2, June 1997

Zhang, Y.-J., Murray, M. & L. Ferreira (1998a) "An integrated model for track degradation prediction" *8th WCTR proceedings*, Vol 3 pp. 527-539

Zhang, Y.-J., Murray M. & L. Ferreira (1998b) "Effects of substructure on track performance"; *Traffic and Transportation studies, proceedings of the ICTTS*, July 27-29 1998, pp. 850-859

Zakharov, S. & I. Zharov (2002) "Simulation of mutual wheel/rail wear"; *Wear 253*, pp. 100-106

Zicha, J.H. (1989) "High speed rail track design" *Journal of transportation engineering*, Vol 115 No.1, p.68-83, American Society of Civil Engineers

Zoeteman, A. (2004) "Railway Design and Maintenance from a Life-Cycle Cost Perspective" *PhD thesis* Delft University of Technology, Netherlands

Zoeteman, A. & P. Veit (2003) "Ten years of Life Cycle Costing in Austria and the Netherlands: An overview of the performed R&D and an agenda for the future", *Proceedings of the 6th World Conference on Railway Research 2004*, Edinburgh UK

10.2 SBB regulations

R RTE 21110	Unterbau und Schotter, Normalspur (und Meterspur)
R RTE 220.41	Lückenlose Gleise, lückenlos verschweisste Weichen und verlaschte Gleise, Normalspur
R 220.66	Einbau, Kontrollen und Unterhalt von Weichen
R I-22067	Einbau, Kontrollen und Unterhalt von Schnellfahrweichen
R I-22220	Ultraschallprüfungen von Schienen und Weichenbauteilen
R RTE 22240	Schweissarbeiten an Schienen und Weichenbauteilen
Weisung W BT 24/97 (Fb/Ob 96)	Materialausbrüche an Weichenbauteilen
D I-EB-FB 02/2004	Anleitung zum Gebrauch der Kontrollehren für die Prüfung der Entgleisungssicherheit in Weichen

Annexes

Annex 1 Discontinuous turnouts: stub turnouts 113
Annex 2 Formulas for Evolution of Dynamic Impact Factors for Vertical Wheel Loads Acting on Rail 115
Annex 3 S&C degradation and deterioration. 117
Annex 4 Database description: Switches & crossings 121
Annex 5 Database description: S&C maintenance 123
Annex 6 Database description: S&C renewals 125
Annex 7 S&C component renewal database 127
Annex 8 S&C maintenance work modelling 129

Annexes

Annex 1 Discontinuous turnouts: stub turnouts

On older railway or tramway networks, sometimes turnouts can be found with a continuous guiding but always only in one direction. These devices do not use point rails, but stub rails, one replacing the other when changing the direction where the turnout leads. Because of the fact that if they are used in the false direction a derailment occurs, they are never used anymore in modern rail or tramways. The pictures below are only for explanatory purposes.

Source: Müller, C. "Der Totalumbau der Pöstlibergbahn", Der Eisenbahningenieur, Oktober 2007 p. 49-53

Annex 2 Formulas for Evolution of Dynamic Impact Factors for Vertical Wheel Loads Acting on Rail

In chapter 3 of this report the following figure is used to explain dynamic train loads.

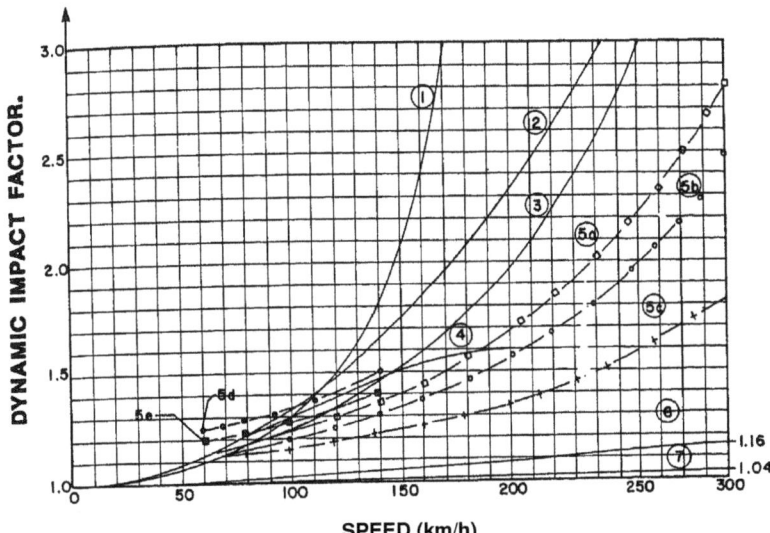

The sources of the curves are the following:

Curve 1 – Winkler's first formula

$$\eta = 1 \Big/ \left[1 - M_0 \left(v^2 / EIg\right)\right], \text{ where } M_0 = 0.1888 \; Pd \quad \text{[Ref 1]}$$

P = wheel load in kg*/cm² (1kg* = 10N) and d is spacing of ties in cm.

Curve 2 – Formula according to former Association of Central European Railroads (1936)

$$\eta = 1 + \left(v^2 / 30.000\right)$$

Curve 3 – Winkler's second formula for $\bar{v} \leq 65 \text{ km/h}$

$$\eta = 1 \Big/ \left[\left(1 - 7 \times 10^{-8} v^2\right) \times (Pd/J)\right] \quad \text{[Ref 1]}$$

Klika's formula for $\bar{v} > 65 \text{ km/h}$

$$\eta = 1 \Big/ \left[1 - \left(9.1 \times 10^{-6} v - 2.957 \times 10^{-4}\right) \times (Pd/J)\right] \quad \text{[Ref 2]}$$

Annexes

Curve 4 – Schramm's formula

$\eta = 1 + (4.5v^2/100.000) - (1.5v^3/10.000)$, where v is in kph [Ref 5]

Curve 5 – Birman's formulas

$\eta = 1 + \alpha + \beta + \gamma$, $\alpha = 0.04\,(v/100)^3$, $\eta = 0.2$, $\gamma = \gamma_0 \times ab$, $\gamma_0 = 0.1 + 0.017\,(v/100)^3$ [Ref 4]

Curve 6 – Actual impact factors from testing of TGV 001 vehicle [Ref 3]

Curve 7 – Theoretical formula for ideal track and vehicle without irregularities where $\beta = \sqrt[4]{K/4EJ}$, $u = $ mass in $\text{kg} * \text{s}^2 \times \text{cm}^{-1}$, $\omega_0 = \sqrt{k/u}$,

$\eta = 1 + (1/2) \times (c^2 \beta^2 / \omega_0^2)$, $K = $ elasticity modulus of track $\times$ tie width

References:
1. Winkler, E. (1871). Vortrage über Eisenbahnbau. Band 1: Der Eisenbahn – Oberbau. 2 Aufl. H Dominicus, Praha, Czechoslovakia (in German).
2. Klika, E. (1948). "Studie o novych zasadach vypoctu napeti v kolejnicich." Technicky Obzor, Prague, Czechoslovakia, 56 (6), (in Czech).
3. Erchkov, O. P., and Kartzev, V. J. (1980). "Recherches théoriques et expérimentales sur les mouvements des véhicules ferroviaires circulant a une vitesse de 200 km/h et exigence relatives a l'entretien des lignes a grandes vitesses." Rail Int., 714 – 720 (in French).
4. Birman, F. (1966). "Track parameters static and dynamic." Proc., Inst. Mech. Engrs.
5. Schramm, G. (1955). "Die Beanspruchung der Schienen durch die Eisenbahn-Fahrzeuge." Glases Annalen 79 (11).

Source:

Zicha, J.H. (1989) "High speed rail track design" *Journal of transportation engineering,* Vol 115 No.1, p.68-83, American Society of Civil Engineers

Annex 3 S&C degradation and deterioration.

This appendix contains an example of a list with mechanical switch failures and deterioration processes. They are mainly based on a 1:9 standard switch with a constructed frog on wooden sleepers. The failures for a switch on concrete sleepers or on concrete or wooden sleepers with a moveable frog or cast manganese frog can be different for some wear phenomena.

Switch engine (actuator) failures and switch heating failures are not included.

Explanation of the different columns

#
Arbitrary numbering of different failures

Part
Description of various parts of a switch which can show a technical / mechanical defect

Possible techn. def. / failure / degrad.
Description of the type of failure on every S&C component mentioned in the previous column.
- techn. = technical
- def. = defect
- degrad. = degradation

Possible cause
A reason which might have caused the occurrence of the failure.

Maintenance action
The works that have to be carried out to repair the switch or crossing, or to restore the required situation.

Current insp.
The inspection method during which the failure is detected nowadays.
insp. = inspection
US = Ultrasonic inspection
VI = Visual Inspection

Number 1 and 2 and the last two are geometrical problems. The rest are regarded as material failures.

Annexes

#	Part	Possible techn. defect / failure / degradation	Possible causes	Maintenance action	Current inspection method
1	Complete switch (geometry)	Switch subsided	Subsiding (sleeper-ballast interaction)	Tamp	VI
2	Complete switch (geometry)	Switch shifted	Shift (sleeper-ballast interaction)	Tamp and shift	VI
3	Rails	Gauge problem	Free motion, excavation, wear on fastenings, sleeper wear	Restore gauge (base plates)	VI
4	Rail	Rail crack	Wear	Replace/renew rail	VI
5	Rail	Rail crack	Wear	Replace/renew rail	Ultrasonic inspection
6	Rail	Broken rail	Wear	Temporarily: attach emergency fish plate; Final: replace rail	-
7	Rail	Rail tear vertically (railhead height)	Tear	Replace/renew rail	(occurs almost never)
8	Rail	Rail tear lateral direction	Tear	Replace/renew rail	VI
9	Rail	Rail wave defects	Tear	Grind rail	VI / measure
10	Rail	Rail wave defects	Tear	Grind rail	Ultrasonic
11	Rail	Rail defect (e.g. head checks)	Wear due to rolling contact	Grind rail	VI
12	Rail	Rail damage (bullet holes, burned spots)	Vandalism, wheel spin	Grind/replace rail	
13	Rail	Base plate worn out due to rail foot movement	Tear due to movements on the baseplate	Tighten fixings	VI
14	Electrical connection to rail	Electrical connection broken	Vandalism, work on switches	Repair	VI
15	Metallurgic weld	Cracked / broken weld	Wear, slag inclusion	Thermit weld: renew completely; Flash but and arc weld: replace	VI
16	Metallurgic weld	Cracked / broken weld		Thermit weld: renew completely; Flash but and arc weld: replace	US inspection
17	Metallurgic weld	Sagged / bended weld	Subsiding / lateral displacement	Straighten weld	VI
18	Insulated joint	Sagged / bended weld	Burrs, grindings	Grind burrs, remove grindings carefully	VI
19	Insulated joint	No electrical isolation	Contamination in weld	Replace with constructed joint	VI
20	Insulated joint	Insufficient electrical isolation	Movement in weld		VI
21	Insulated joint	Gap between rail ends	Subsiding / lateral displacement	Straighten weld	-
22	Insulated joint	Sagged / bended weld	Ageing / movement due to traffic load	Fill space	VI
23	Isolation plate	Crack in plate/ plate tear	Subsiding / lateral displacement	With constructed joint replace; glued joint install prefab joint	-
24	Non isolating joint	Sagged / bended weld	Subsiding / lateral displacement	With constructed joint replace, glued joint install prefab joint	-
25	Conducting glued joint	Sagged / bended weld	Wear	Straighten weld	VI
26	Fish plate	Broken / cracked fishplate	Vibrations	Constructed joint: tighten or apply; glued joint: install prefab joint	VI
27	Nuts and bolts	Nuts and bolts are loose / lost	Wear	Replace rigid clip	VI
28	Rigid clip	Broken rigid clip	Tear	Replace/tighten rigid clip	VI
29	Rigid clip	Rigid clip is worn	Vibrations, tightened too much	Tighten fastening or replace nuts and spring rings	VI
30	Nuts and spring rings	Loose bolt, broken bolt	Ageing	Check base plates for wear	VI
31	Base plates	Baseplate is tamped into sleeper	Vibrations	Refit baseplate, check gauge	-
32	Coach screws	Coach screw untight	Tightened too much	Fill old hole	-
33	Coach screws	Stripped coach screw	Tightened too much	Place new coach screw with coil or move sleeper	VI
34	Coach screws	Broken coach screw	Derailment, damage through tamping	Replace sleeper	VI
35	Wooden sleeper	Broken or cracked sleeper	Ageing, overload	Move sleeper and make new holes / re-glue	VI
36	Wooden sleeper	(Slide) baseplate is tamped into sleeper	Loose attachment, loose coach screw	Renew / tighten fixing	-
37	Sleeper anchors	Loose anchor	Object in switch flangeway	Remove object	-
38	Half set of switches	Switch doesn't move	Ballast in switch flangeway	Check on ballast excavation, excavate ballast	-
39	Half set of switches	Switch doesn't move	Various causes, increased friction	Re-adjust/grease	-
40	Half set of switches	Switch movement too slow	Burrs, damage	Re-adjust	VI
41	Switch rail	Gap between stock rail and closed switchrail	Switch rail deformation rail hit by trains	Re-adjust	VI
42	Switch rail	Switch flangeway too narrow	Wear	Replace half set of switches	-
43	Switch rail	Broken switch rail / crack lengthwise	Wear	Carry out US measurements	Ultrasonic
44	Switch rail	Broken switch rail / crack lengthwise	Flattened by passing trains	Grind burrs, after that re-adjust drive and control rods	VI
45	Switch rail	Burrs on switch rail	Tear	Grind switch rail laterally, or replace switch rail	VI
46	Switch rail	Switch rail lateral tear	Tear	Grind switch rail, top of switch rail	VI
47	Switch toe	Too sharp switch rail / shelling	Tear caused by trains	Replace half set of switches	VI
48	Switch toe	Switch toe collided by train	Too much tear on stock rail	Grind switch rail	VI
49	Switch rail baseplate	Switch toe lies too high	Track maintenance	Replace switch rail baseplates	VI
50	Stock rail near switch rail	Plates broken or damaged Stock rail worn laterally	Tear	Replace half set of switches	VI

Annexes

#	Component	Defect	Cause	Action	Type
51	Stock rail near switch rail	Burrs on stock rail	Flattened by passing trains	Grind stock rail	VI
52	Slide baseplate	Broken or cracked slide baseplate	Wear	Replace slide baseplate	VI
53	Baseplate to sleeper bolts	Bolts loose or broken	Wear, vibrations	Tighten or replace bolts	VI
54	Lift plate	Bolts loose or broken	Wear, vibrations	Tighten or replace bolts	VI
55	Switchglide	Switchglide broken / cracked	Wear	Replace Switchglide	VI
56	Switchglide	Switchglide contaminated / torn	Pollution, tear	Clean switchglide	VI
57	Steel role at toe	Role jammed	No / insufficient greasing	Remove old grease and contamination	1 week / 3 weeks
58	Synthetic role at toe	Role jammed or torn	Tear	Replace role	VI
59	Synthetic role at toe	Role jammed or torn	Pollution	Clean role	6 months
60	Frog	Frog cracked	Wear, hit by train, material failure	Grind and/or weld a new piece on or replace completely	VI
61	Frog	Frog cracked		Carry out US-inspection	US-inspection
62	Frog	Frog damaged horizontally	Hit by train	Replace frog	VI
63	Frog	Frog damaged horizontally	Hit by train	Replace frog	VI
64	Frog	Frog torn vertically	Tear	Weld new small plates on	VI
65	Frog	Burrs on frog	Flattened by passing trains	Grind	VI
66	Frog	Shelling on frog	Tear	Grind, weld new plates on or replace	VI
67	Horizontal bolts	Loose / broken bolts	Wear / vibrations	Tighten or replace bolts	VI
68	Check rail	Check rail torn	Hit by train	Adjust flangeway by adding or removing fill plates	VI
69	Bolt check rail to sleeper	Loose / broken bolt	Wear, damage	Tighten or replace bolt	VI
70	Ballast	Subsidence of the track	Subsidence	Renew ballast	VI
71	Ballast	Ballast is conducting	Ageing, ballast contaminated	Renew ballast	-

Source: Jovanović, S. & W.-J. Zwanenburg (2002) "Switches and Crossings Management System: EcoSwitch – Feasibility Study" *ERRI Project Report D251/RP1* European Rail Research Institute, Utrecht, the Netherlands

Annex 4 Database description: Switches & crossings

This annex describes all fields in the database WEITABL1 as composed by the SBB in March 2005. The fields provide a complete description of all switches and crossings on the SBB network.

FIELD	Contents	Value	Meaning of value
A0151_NR	Internal number in DfA-database		
BP_S	Abbreviation of station or node *Betriebspunkt*		
W_NR	Switch or crossing number		
W_NRA	Alfabetic attribution to W_NR		
W_Z	Status	1	Existing
		3	To be demolished
		4	Project
		5	Temporarily
W_ART	Geometrical shape	EW	Standard turnout
		SW	Symmetrical turnout
		DW	Combined turnout
		EKW	Diamond crossing with single slip
		DKW	Diamond crossing with double slip
		GD	Diamond crossing
		MS	Centre piece of a double intersecting cross-over
		ZV	Protection switch or switch rail
		AW	Narrow gauge entrance or exit switch in normal gauge track
SIPR	Rail profile	I	SBB I (46 kg/m)
		IV	SBB IV (54 kg/m) = UIC 54 E =54E2
		VI	SBB VI (69 kg/m) = UIC 60 = 60E1
		V	SBB V (36 kg/m)
W_TYP	Switch or crossing type by radius		160, 185, 300, 500, 900, 1600 .....
W_LAGE	Curved or not	*	Unknown
		G	Basic form
		B	Curved switch or crossing
W_NEIG	Switch or crossing type by angle		1:9, 1:12, 1:14, ... 2*(1:9) ...
SCHW_ART	Sleeper type	H	Wood
		Be	Concrete
		S	Steel (rails not isolated)
		Si	Steel (electrically isolated rails))
		O	Without sleepers
		HEA	Steel beam support type HEA
		HEB	Steel beam support type HEB
ABLK	Switch or crossings direction (left-hand switch etc.)		L, R, LL, LR, RL, RR
ZUNGEN	Switch rail type	F	Spring switch rail
		G	Articulated switch rails
		FT	Spring switch rail with damper.
GL_NR	Track number		
GLEISKAT	Track categorie	ANG	Connection to private tracks
		HG1	Main / principal track 1
		HG2	Main / principal track 2
		HG3	Main / principal track 3
		NG1	Secondary track 1
		NG2	Secondary track 2
		NG3	Secondary track 3
		KTU	Other Swiss or foreign owner
J	Inspection and check-up category		Every 2, 4, or 6 years
V_PLAN	Number of lay-out plan		
TP_NR	Number of type plan		
TP_ZUS	Attribution to TP_NR		
TP_IND	Change index of TP_NR		
SPURW	Track gauge	N	Normal gauge
		M	Narrow gauge (1000mm)
		3S	3-rail track
		4S	4-rail track
		S	Special track
LG_STM	Length in main direction		[m]
LG_ABL	Length in diverting direction		[m]
WLAGE	Curved or not		Curved / not-curved
RMIN	Mimimal radius in main direction		[m]
UMAX	Cant in main direction		[mm]
JAHR	Year of placements		
G	Way of placement	G1	New rods, new sleepers
		G2	Old rods, old sleepers
		G4	Old rods, new sleepers
		GX	Unknown
VORHER	Duration previous switchr or crossing was in place		[Years]

FIELD	Contents	Value	Meaning of value
BEMKG1	Remarks1		
BEMKG2	Remarks2		
RESERVE1	Stock of special components 1		
ZEICHNR1	Drawingnumber of 1st special component		
LAGER1	Place where 1st special compenent is stocked		
RESERVE2	Stock of special components 2		
ZEICHNR2	Drawingnumber of 2nd special component		
LAGER2	Place where 2nd special compenent is stocked		
TEILE	With or without special S&C components	N	Normal design
		S	Special design
SKAP	Mit oder ohne Schwellenkappen	K	With sleeper covers
		O	Without sleeper covers
S_KAPPEN	Information on the sleeper covers		
SCHOTTER	Ballast type	A	Regenerated used ballast
		F	Small grain ballast
		N	Normal ballast
		O	Without ballast
		R	Ballast of broken cobles
		1	Quality 1
		2	Quality 2
		3	Quality 3
SCHWVERF	Welding type	A	aluminothermic.
		E	Electric
		0	Fish-plated
SCHWDAT	Welding date		
SCHITEMP	Rail temperature before welding		
IMPRESA	Welding company		
UNTJAHR	Renovation of substructure		
UBDICKE	Substructure thickness		[cm]
BEMKG3	Remarks on substructure 1		
BEMKG4	Remarks on substructure 2		
W_HEIZG	Switch rail heating	E16	electric 16 2/3 Hz
		E50	electric 50 Hz
		GAS	Gas: national network
		O	without heating
		PBH	Propane, Container
		PZT	Propane, central tank
		P30	Propane, Tank: 300 kg
		WW	How water
		EM	el. BLS Mittelland
		EO	el. BLS Oberland
WH_FNST	Remote control of switch heating	F	with remote control
		O	without remote control
ANTR	Switch motor / actuator	1	unknown
		2	elektric
		3	mechanical
		4	without
ZENT	Remote control of switch actuator / motor	1	unknown
		2	centralised
		3	not centralised
VSCH_ANZ	Switch rail locking number		
VERS	Switch rail locking type	1	unknown
		2	different
		3	Klinkenverschlüsse
		4	Jüdel
ISOL	Electric isolation of the rails	1	unknown
		2	isolated
		3	not isolated.
EGTM	For switches connecting to in private tracks: part of ownership		[%]
UNTH	For switches connecting to in private tracks: part of maintenance		[%]
BEMKG5	For switches connecting to in private tracks: other info		
BEMKG6	For switches connecting to in private tracks: other info		
NFGDAT	Date indicating database entry		
G_JAHR_	Proposed renewal		proposed year
SU_JHR_	Next planned systematic maintenance (tamping)		proposed year
SLS_JH_	Next planned screw hole renovation action		proposed year
AUFTSJ_	Next planned welding-on action		proposed year
SCHL_J_	Mext planned grinding action		proposed year
KONTRJ_	Next switch inspection / check-up		proposed year
BEM_VA_	Remark with planned works.		

Annex 5 Database description: S&C maintenance

The database called WEITABL2, provided in March 2005 by SBB contained a summary of all maintenance works. Every record is one work for which the field A0151_NR forms the connection to the switch or crossing on which the maintenance work is carried out.

FIELD	Contents	Value	Meaning of value / remark
A0151_NR	Internal number in DfA-database		
BP_S	Abbreviation of station or node *Betriebspunkt*		
W_NR	Switch or crossing number		
W_NRA	Alfabetic attribution to W_NR		
W_Z	Status	1	Existing
		3	To be demolished
		4	Project
		5	Temporarily
JHR	year the work is carried out		
UMB	New construction or renewal		free text
SU	systematic maintenance (mainly tamping)		free text
SLS	Screw hole renovation with synthetic raisin		free text
ASW	Welding (on the frog)		free text
SCL	Grinding		free text
KON	Control incpection		free text
BEM	Remarks		

Annex 6 Database description: S&C renewals

WEITABL3 contains the information on:
- replaced switch rail with accompanying stock rail,
- frog
- common crossing
- moveable frogs
- check rail

A distinction is made with regard to which switch rail with accompanying stock rail is replaced. LL means in this case: was the left

FIELD	Contents	Value	Meaning of value / remark
A0151_NR	Internal number in DfA-database		
BP_S	Abbreviation of station or node *Betriebspunkt*		
W_NR	Switch or crossing number		
W_NRA	Alfabetic attribution to W_NR		
W_Z	Status	1	Existing
		3	To be demolished
		4	Project
		5	Temporarily
JHR	year the work is carried out		
Z_LL	Number of replaced switch rail with accompanying stock rail LL		
Z_LR	Number of replaced switch rail with accompanying stock rail LR		
Z_RL	Number of replaced switch rail with accompanying stock rail RL		
Z_RR	Number of replaced switch rail with accompanying stock rail RR		
H_EF	Number of replaced frogs		
H_DP	Number of replaced common crossings		
H_BW	Number of replaced moveable frogs		
R_LL	Number of replaced check rail LL		
R_LR	Number of replaced check rail LR		
R_RL	Number of replaced check rail RL		
R_RR	Number of replaced check rail RR		
SCW	Number of replaced sleepers		
BEM	Remarks g		

Annex 7 S&C component renewal database

Data analysis

For the following components renewals, sufficient data is available in the database WEITABL3 with 17.171 entries.

- Half set of switches and accompanying stock rail
    o divided into 4 different directions: LL, LR, RL, RR
- Check rail
    o Divided into 4 different directions: LL, LR, RL, RR
- Frog
    o Divided into a normal frog, a common crossing or a moveable frog.

The first entrances in the database date from 1968. But it is confirmed that only since the 1990's the database has been updated consistently and can be regarded as complete.

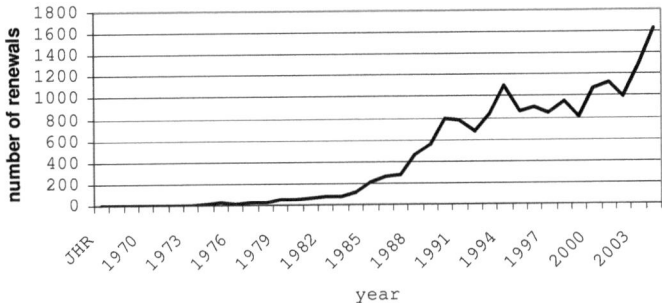

Figure 64 Renewals per year as registered in the database WEITABL3

Annex 8 S&C maintenance work modelling

Data analysis
In database WEITABL2 with a total of 114,375 registrations over the period 1900-2004, but most of them are from last years as the figure shows

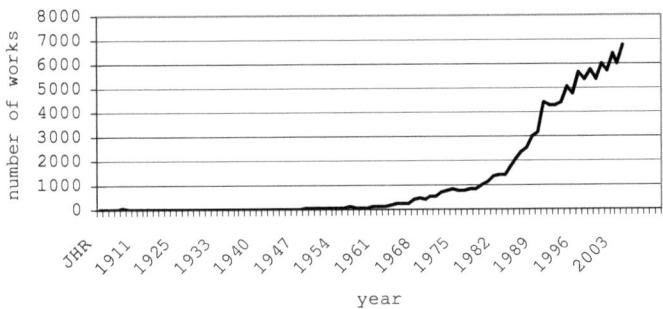

Figure 65 Works per year as registered in the database WEITABL2

The following works can be found in the database:

- Big renewal of multiple parts (or renewal with used parts), but not an integral renewal
- Systematic maintenance (mainly tamping)
- Screw hole renovation with synthetic raisin
- Welding (on the frog)
- Grinding
- Control inspection

Big renewals of multiple parts include
- the integration of a switch or crossing in the continuous welded rails of surrounding tracks. This is a unique work and not related to wear or degradation,
- most of the partial renewals as mentioned here are with used parts are on secondary tracks and therefore also not part of this project, which focuses on main tracks, where renewals take place because of degradation and wear.

The screw hole renovation is a unique work only carried out on secondary tracks on switches and crossings on wooden sleepers. This renovation is also not based on a wear or degradation regime, but more to increase the service life of switches and crossings on wooden sleepers.

The control inspection (frequency) is prescribed by regulation and not related to a wear regime.

This means this analyses will only focus on the systematic maintenance, from now on: called tamping, the grinding and the welding on the frog.

All pictures are from the author or from LITEP unless otherwise indicated.

This report has been written in Word 2002, 2003 and 2007.
The databases have been established with Microsoft Access 2002 and 2007.
The graphs and statistical analyses and models have been produced with Matlab R2007b and R2008b.

The text has been consolidated Wednesday 3rd of December 2008.

Latest changes have been made for this version the 19th of March 2009.

Die VDM Verlagsservicegesellschaft sucht für wissenschaftliche Verlage abgeschlossene und herausragende

Dissertationen, Habilitationen, Diplomarbeiten, Master Theses, Magisterarbeiten usw.

für die kostenlose Publikation als Fachbuch.

Sie verfügen über eine Arbeit, die hohen inhaltlichen und formalen Ansprüchen genügt, und haben Interesse an einer honorarvergüteten Publikation?

Dann senden Sie bitte erste Informationen über sich und Ihre Arbeit per Email an *info@vdm-vsg.de*.

Sie erhalten kurzfristig unser Feedback!

VDM Verlagsservicegesellschaft mbH
Dudweiler Landstr. 99   Telefon +49 681 3720 174
D - 66123 Saarbrücken   Fax   +49 681 3720 1749
www.vdm-vsg.de

Die VDM Verlagsservicegesellschaft mbH vertritt

Printed by Books on Demand GmbH, Norderstedt / Germany